ADVANCES IN SPACE RESEARCH

The Official Journal of the Committee on Space Research (COSPAR)
A Scientific Committee of the International Council of Scientific Unions (ICSU)

VOLUME 18, NUMBER 7

SATELLITE DATA FOR ATMOSPHERE, CONTINENT AND OCEAN RESEARCH

Meeting A4

Sponsors

SCIENTIFIC COMMITTEE ON OCEANIC RESEARCH (SCOR)
INTERNATIONAL ASSOCIATION FOR THE PHYSICAL SCIENCES OF THE OCEAN (IUGG/IAPSO)

Program Committee

F. Wilkerson, U.S.A.
R. C. Dugdale, U.S.A.
A. Morel, IUGG/IAPSO Representative
M. Fassham, U.K.
P. Schlittenhardt, Italy
J-P. Malingreau, France
T. Platt, SCOR Representative

SATELLITE DATA FOR ATMOSPHERE, CONTINENT AND OCEAN RESEARCH

Proceedings of the A3 and A4 Meetings of COSPAR Scientific Commission A which were held during the Thirtieth COSPAR Scientific Assembly, Hamburg, Germany, 11–21 July 1994

Edited by

E. RASCHKE

Institute for Atmospheric Physics, GKSS Research Center, D-21502, Geesthacht, Germany

V. V. SALOMONSON

NASA Goddard Space Flight Center, Eark Sciences Directorate, Greenbelt, MD 20771, U.S.A.

R. STÜHLMANN

Institute for Atmospheric Physics, GKSS Research Center, D-21502, Geesthacht, Germany

and

F. WILKERSON

University of Southern California, Los Angeles, CA 90089-0371, U.S.A.

Published for

THE COMMITTEE ON SPACE RESEARCH

PERGAMON

U.K.	Elsevier Science Ltd, The Boulevard, Langford Lane, Kidlington, Oxford OX5 1GB, England
U.S.A.	Elsevier Science Inc., 660 White Plains Road, Tarrytown, New York 10591-5153, U.S.A.
JAPAN	Elsevier Science Japan, Tsunashima Building Annex, 3-20-12 Yushima, Bunkyo-ku, Tokyo 113, Japan

First edition 1995

ISBN 0-08-042672-7

In order to make this volume available as economically and as rapidly as possible the author's typescript has been reproduced in its original form. This method unfortunately has its typographical limitations but it is hoped that they in no way distract the reader.

NOTICE TO READERS

If your library is not already a subscriber to this series, may we recommend that you place a subscription order to receive immediately upon publication all new issues. Should you find that these issues no longer serve your needs your order can be cancelled at any time without notice. All these conference proceedings issues are also available separately to non-subscribers. Write to your nearest Elsevier Science office for further details.

Transferred to digital print 2008
Printed and bound in Great Britain by
CPI Antony Rowe, Chippenham and Eastbourne

CONTENTS

Chapter 2 – OCEAN AND LAND PRODUCTIVITY AS DERIVED FROM SATELLITE DATA (Mtg A4)

Chapter 1

**Estimation of Moisture in the Atmosphere and in the Ground
from Satellite Data (Mtg A3)**

SUMMARY

Estimation of Moisture in the Atmosphere and in the Ground from Satellite Data

Life needs water. Moreover, there is a manifold of non-linear interactions in our climate system, where water in all of its three phases plays a dominant rôle. Water transports energy and many pollutants; water in the atmosphere modifies very effectively the energy transports by radiation, which in turn force many dynamical processes.

Therefore, we must know quite accurately the energy and water transports in our climate system - i.e. within the atmosphere, at ground, and exchanges between them - with quite high regional and temporal detail. And furthermore, numerical models for weather forecast, for climate simulations, for estimate of transports of additional materials - in particular pollutants - and simulations of the impact of climate variations must be capable to simulate such transports with realistic detail for the present time and the future. These ambitious goals are well defined within the scope of the GEWEX, Global Energy and Water Cycle Experiment, a subprogramme of the World Climate Research Programme.

GEWEX has a major observational component, to measure and monitor with space-based instruments the atmospheric water vapour in all layers of the troposphere, the cloud water contents and the partition of its liquid and solid phases, the precipitation over continents and oceans - also in form of snow - and also various surface characteristics, which allow to estimate the evaporation and evapotranspiration of water.

This symposium reviewed in 13 papers, of which finally 12 are available for publication, several methods to estimate such water transports from the presently available satellite data sets. It unfortunately missed the most recent multispectral imaging spaceborn radar measurements.

 Pergamon

Adv. Space Res. Vol. 18, No. 7, pp. (7)5–(7)16, 1996
Copyright © 1995 COSPAR
Printed in Great Britain. All rights reserved
0273–1177/96 $9.50 + 0.00

0273–1177(95)00282–0

ATMOSPHERIC WATER VAPOUR AND CLOUD WATER: AN OVERVIEW

E. Ruprecht

Institut für Meereskunde, Düsternbrooker Weg 20, 24105 Kiel, Germany

ABSTRACT

Hydro-meteorological parameters i.e. precipitable water, cloud water and ice content, and precipitation are most variable parameters in the atmosphere. This is the main reason why representative direct measurements of these properties are hardly available. Remote sensing with satellite-borne instruments in particular in the microwave spectral range is a way out of this dilemma. A number of algorithms has been developed. The different methods how to proceed in the development of such algorithms are discussed.

Verification of the retrieved products in particular the liquid water path is a great problem, a few ideas will be discussed. Results will be shown for the total precipitable water W and liquid water path LWP over the Atlantic Ocean for different time scales. The structure of the W field is very similar for the same month in different years. But LWP is very variable, even for monthly means (October 1987 and 1989) the differences can be larger than 0.1 kg/m^2.

1. INTRODUCTION

Water vapour is the most important constituent of the atmosphere for all weather and climate processes. It is the only gas which can exist in all three phases at the temperature and pressure range of the atmosphere. In the gaseous phase it is very active as greenhouse gas; it strongly affects the radiation budget by cloud particles (liquid and solid), and it provides the necessary precipitation for the development of vegetation. In addition to the influence on the energy budget by its radiative effects, the phase transition is for many phenomena the main local energy source or sink.

Our knowledge about the global distribution of the water in the atmosphere stands in contrast to its importance. This is due to the limited observation methods and the high spatial and temporal variability of this atmospheric component. Radiosonde measurements are operationally used for the observations of water vapour. Their limitations are well known (e.g. Nash and Schmidlin, 1987). There exists, however, no operational method to measure cloud and precipitation water in the atmosphere.

Because of the radiative effects of atmospheric water it seems logical to use space-borne radiometers for its retrieval. Observations in all spectral ranges are applied to estimate total precipitable water, humidity profiles, cloud liquid and ice water content, and precipitation.

For the estimation of total precipitable water methods make use of thermal IR observations in the window from 8 to 13 μm (split window methods) where the water vapour continuum mainly determines the radiances (Schmetz and van de Berg, 1991) or of microwave observations in the centre and the wings of different water vapour absorption lines (e.g. for SSM/I (Special Sensor Microwave/Imager), Alishouse et al., 1990, Schlüssel and Emery, 1990, Fuhrhop and Ruprecht, 1994). Infrared observations are also used for the retrieval of humidity profiles (e.g. Smith, 1983). Kakar (1983) proposed to apply measurements at the 183 GHz water vapour line to retrieve vertical distribution of moisture, and this is now planned for AMSU (Advanced Microwave Sounding Unit). A very different method was developed by Wagner et al. (1990) in combining statistical information of the humidity profiles (Empirical Orthogonal Functions, EOF) derived from radiosonde data and direct observations in the microwave spectral range. The latter two methods are now combined to develop a new effective retrieval algorithm for AMSU.

Indirect and direct methods are used to derive cloud water. Kriebel (1989) applied AVHRR (Advanced Very High Resolution Radiometer) channel 1 (0.58-0.68 μm) data to determine indirectly liquid water path of water clouds. He made use of Stephens parametrization which relates reflectances of water clouds to their optical depth and the liquid water path to the optical depth (Stephens, 1978). Direct methods are based on microwave observations, where emission and scattering of hydrometeors determine the radiances measured at satellite level (e.g. for SSM/I, Petty and Katsaros, 1990, Karstens et al., 1994).

Despite of these many algorithms, global distribution of most of the hydro-meteorological parameters are not exactly known. The reason for this fact is the lack of globally applicable algorithms. In the following we shall discuss this problem for the microwave radiometry.

2. PHYSICAL BASIS FOR THE DEVELOPMENT OF ALGORITHMS TO RETRIEVE HYDRO-METEOROLOGICAL PARAMETERS FROM MICROWAVE OBSERVATIONS

The physical processes which produce and modify the radiances received at satellite level are well known and an equation exists for their calculations. In order to solve the inverse problem - given the radiances and retrieve atmospheric parameters - simple solutions are not on hand. In general, there are more dependent variables than measurements. For the brightness temperatures T_B which stand for the radiances in the microwave spectral range, the radiative transfer equation reads when scattering is neglected:

$$
\begin{aligned}
T_B = {}& \varepsilon T_s \exp(-\int_0^{z_t} \alpha(z)\,dz) \\
& + \int_0^{z_t} T(z)\alpha(z)\exp(-\int_z^{z_t}\alpha(z')\,dz')\,dz \\
& + (1-\varepsilon)\exp(-\int_0^{z_t}\alpha(z')\,dz')\int_0^{z_t} T(z)\alpha(z)\exp(-\int_0^{z}\alpha(z')\,dz')\,dz \\
& + (1-\varepsilon)\exp(-2\int_0^{z_t}\alpha(z)\,dz)\,T_{00}
\end{aligned}
\tag{1}
$$

with $\alpha = \alpha(\rho_w, T, p, LWC)$ = volume absorption coefficient
ρ_w = water vapour density, T = air temperature, p = pressure, LWC = cloud water content, ε = surface emissivity, T_S = surface temperature, T_{00} = space temperature ($\approx 2.7K$)

The effects of p_w and LWC on the absorption coefficient are distinctly different for different frequencies around the 22.235 GHz water vapour absorption line (Fig. 1).

In the centre of the line water vapour has the greatest effect; at higher frequencies the influence of LWC more and more dominates.

The contribution of each term in the radiative transfer equation (1) to the brightness temperature T_B changes with the optical thickness of the atmosphere. The calculations are shown in Fig. 2 carried out based on about 1300 radiosonde observations over the Atlantic Ocean (the results are kindly provided by R. Fuhrhop), LWC is parametrized with the method given by Karstens et al. (1994).

Since the signal measured by the space-borne radiometer is composed of contributions from the atmosphere and from the (ocean) surface, parameters which describe the state of both can be retrieved. To solve for these parameters we have, as already mentioned, an underdetermined system

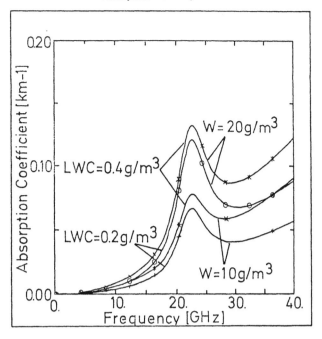

Fig. 1. Volume absorption coefficient around the 22,235 GHz water vapour absorption line in relation to water vapour density and liquid water content (results provided by C. Simmer).

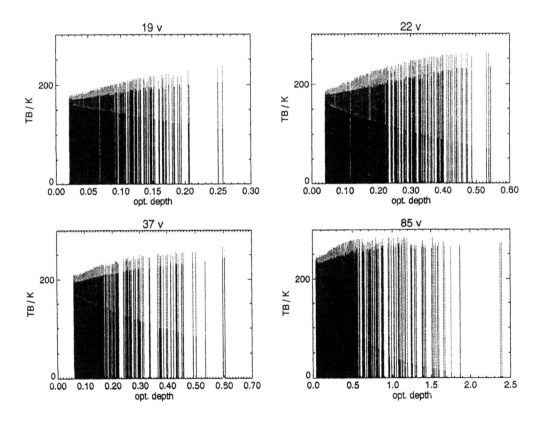

Fig. 2. Contribution of the main terms of equation (1) (first three terms on the right hand side) to the brightness temperature of the four vertically polarized channels of SSM/I in relation to the optical depth of the atmosphere (calculations were carried out by R. Fuhrhop).

black: surface emission, dark grey: atmospheric emission light grey: reflected atmospheric emission

so that for the estimation of W or LWP the linearization can be performed by ln $(T_0 - T_B)$ where T_0 is a reference temperature in general between 280 and 300 K (Chang and Wilheit, 1979). The effect of such linearization is shown for W in Fig. 3 a + b.

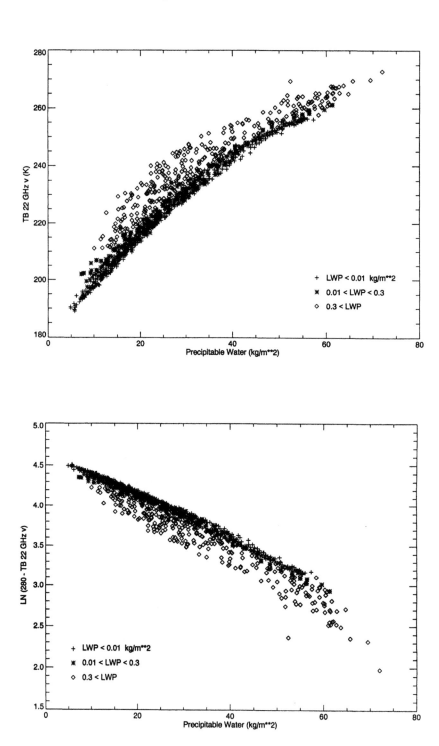

Fig. 3. Simulated 22 GHz brightness temperature in relation to the total precipitable water and liquid water path, LWP.   a) linear abscissa,   b) abscissa: ln(280-T(22V)). (Results provided by R. Fuhrhop).

For the development of an algorithm of either type, physical or statistical, meteorological data are needed to derive empirical coefficients. The success of the algorithm depends on the representativeness of these basic data. If a global set of radiosonde profiles and surface data were available which would include all different weather situations on earth, a global algorithm could be derived which at least in the mean would give correct results. Apparently random errors would nevertheless be generated because there are possible different relations between the atmospheric and surface parameters in different climate regions. This effect we have proved for our precipitable water (W) algorithm. This statistical algorithm is based on about 1300 radiosonde observations over the Atlantic Ocean launched from ships and on the corresponding simulated brightness temperatures:

$$W = 131.95 - 39.50 \ln (280 - T_{22v}) + 12.49 \ln (280 - T_{37v}) \tag{4}$$

The indices of the brightness temperatures give the frequency and the polarization state. The algorithm is validated by additional 250 radiosonde profiles over the Atlantic Ocean the launch times of which coincide with the DMSP (Defence Meteorological Satellite Program, series of satellites which carry the SSM/I) overpass time better than 2 hours. The validation results are given in Fig. 4. The behaviour is typical for most of the published algorithms: overestimation at low W values and underestimation if W increases. The differences are primarily attributed to temperature effects (Fig. 5). If one develops different algorithms for the three temperature ranges given in Fig. 5, large parts of the systematic differences disappear (Fig. 6). The large scatter of the differences in particular for high precipitable water is probably due to cloud effects. We are in the process to work on the problem to decrease this latter scatter.

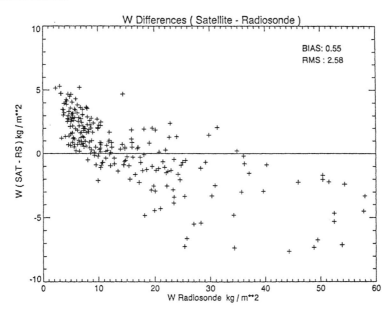

Fig. 4. Difference of total precipitable water retrieved with algorithm of equation (4) and calculated with radiosonde observations against total precipitable water from radiosonde observations.

4. RESULTS

Today we have algorithms available which estimate precipitable water, W, with an accuracy of about 2 kg/m^2. Thus large-scale distribution of W can be derived. Fig. 7 shows an actual case at August 4, 1990 6 UTC. The cyclone over the centre of the North Atlantic Ocean with its dry and wet bands is clearly analysed. High spatial gradients are also found over the western Atlantic which we cannot resolve today with other observation methods. The October 1987 mean is shown in Fig. 8. The striking feature is the almost zonally homogeneous distribution. It is, however, disturbed in those regions where cold waters exist, off the SW and NW coast of Africa or off the E coast of North

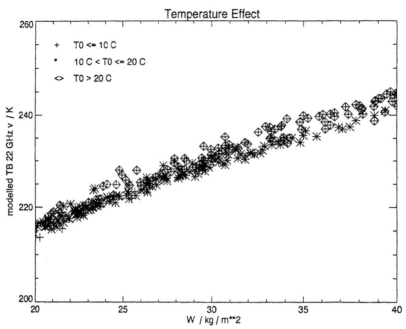

Fig. 5. Simulated 22 GHz brightness temperature against total precipitable water for three low level air temperature ranges. (Fuhrhop and Ruprecht, 1994).

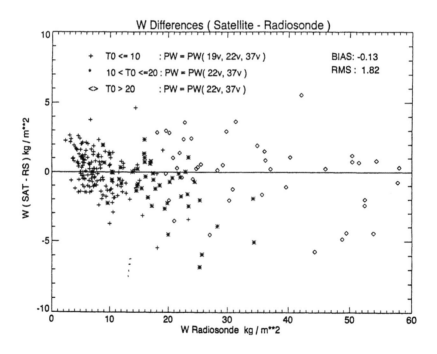

Fig. 6. Difference of total precipitable water retrieved from three different algorithms (one for each temperature range given in Fig. 5) and calculated with radiosonde observations against total precipitable water from radiosonde observations. (Fuhrhop and Ruprecht, 1994)

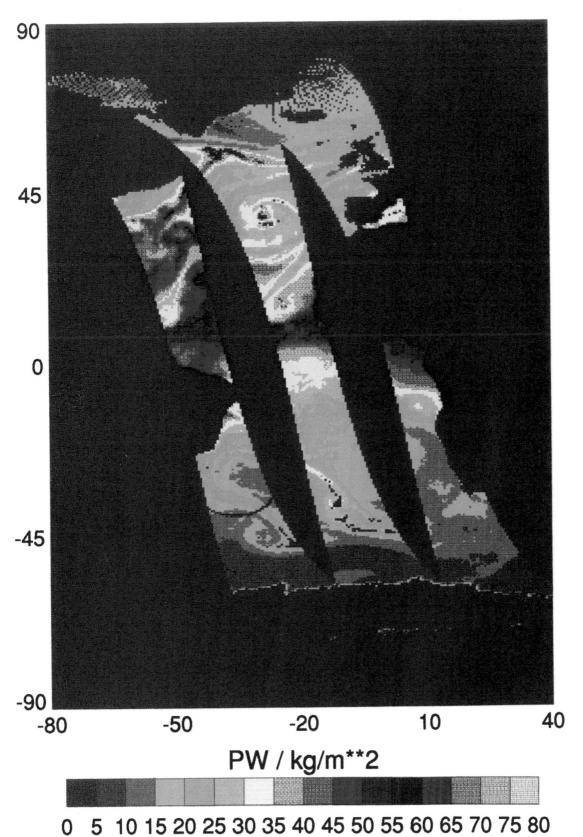

Fig. 7. Total precipitable water derived with equation (4) for the three morning overpasses of DMSP-SSM/I at August 4, 1990.

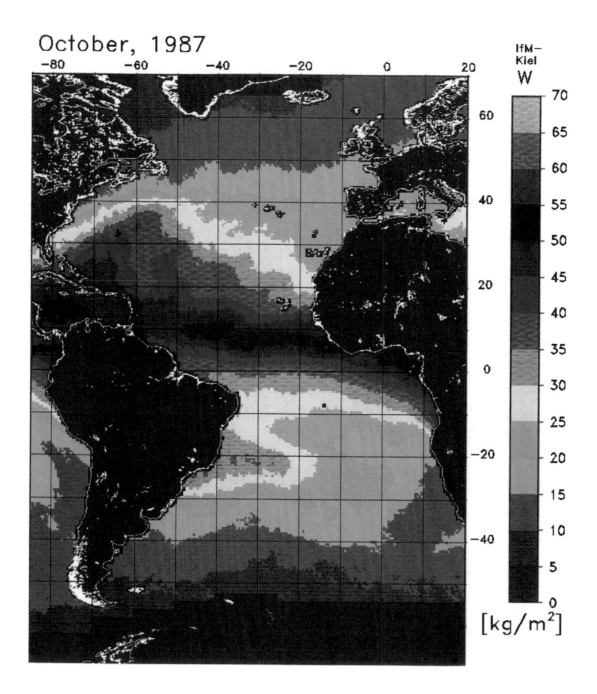

Fig. 8. Monthly mean (October 1987) of total precipitable water derived from SSM/I observations.

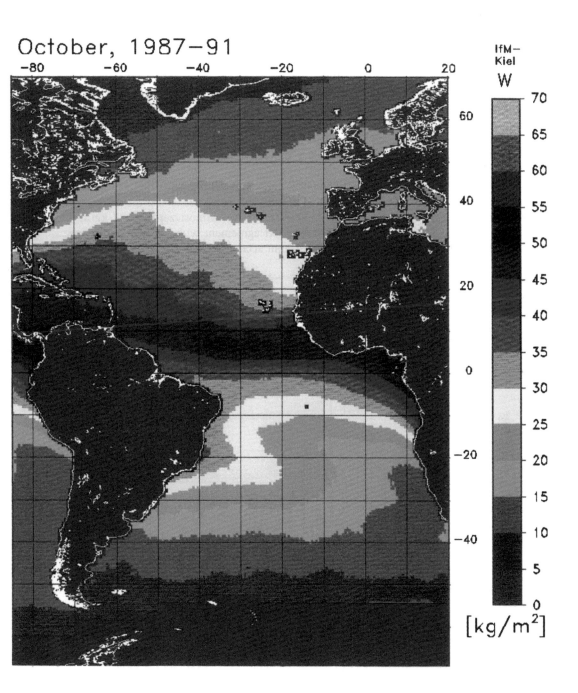

Fig. 9. Longterm October mean (1987-91) of total precipitable water derived from SSM/I observations.

America. Even monthly means over many years are possible to calculate; Fig. 9 gives the five year average of the Octobers from 1987 till 1991. The general structure is very similar to the single year mean of October 1987 but the gradients are weaker.

For the same time periods as above the liquid water path, LWP, of the clouds over the Atlantic Ocean is estimated with an algorithm derived by Karstens et al (1994):

$$LWP = 4.299 - 1.407 \ln(280 - T_{37V}) + 0.400 \ln(280 - T_{22V}) \tag{5}$$

This statistical algorithm is based on simulated brightness temperatures and parametrized LWP. For each radiosonde profile a cloud layer was defined when the relative humidity exceeds 95%. The adiabatic cloud water content was calculated for such a cloud layer and modified according to the results of Warner (1955).

Validation with direct measurements is hardly possible. We did a comparison with the results of a shipborne microwave radiometer which was successful. A very helpful method is to test the algorithm in cloud-free situations. Fig. 10 shows that the mean cloud-free value is almost zero (0.007 kg/m²) and the standard deviation is 0.035 kg/m². The latter is in good agreement with the rms difference from the regression analysis. Thus we estimated for the algorithm accuracy a value of 0.03 kg/m².

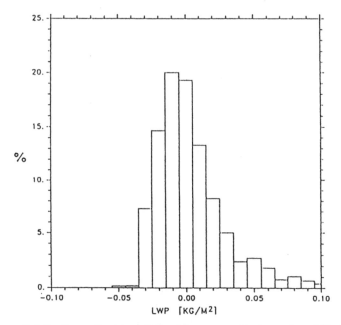

Fig. 10. Frequency distribution of retrieved liquid water path (equation (5)) for cloud free cases (Hargens et al., 1994).

The monthly average of LWP over the Atlantic Ocean is very variable. As an example we show the two months of October 1987 and 1989 (Fig. 11 a,b). Over the Gulf Stream region differences of more than 0.1 kg/m² are apparent, large differences are also found SE of Greenland.

5. CONCLUSIONS

The algorithms to retrieve total precipitable water from microwave observation can be very successfully applied. Their results are helpful to validate the results of weather forecast and climate models in particular in the otherwise datafree regions of the southern oceans. In order to increase the reliability of the algorithm results in these regions, more radiosonde observations are needed in particular for the times of the DMSP overpasses. Thus every opportunity at research vessels etc. should be used.

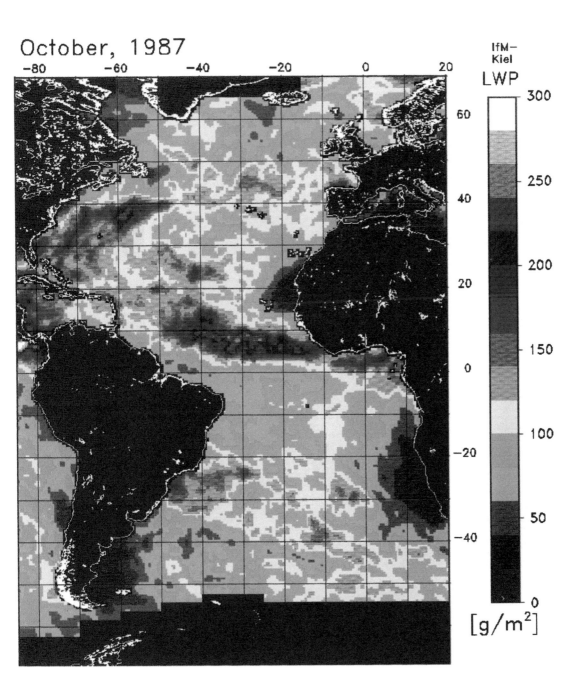

Fig. 11a. Mean monthly liquid water path of clouds over the Atlantic Ocean: October 1987.

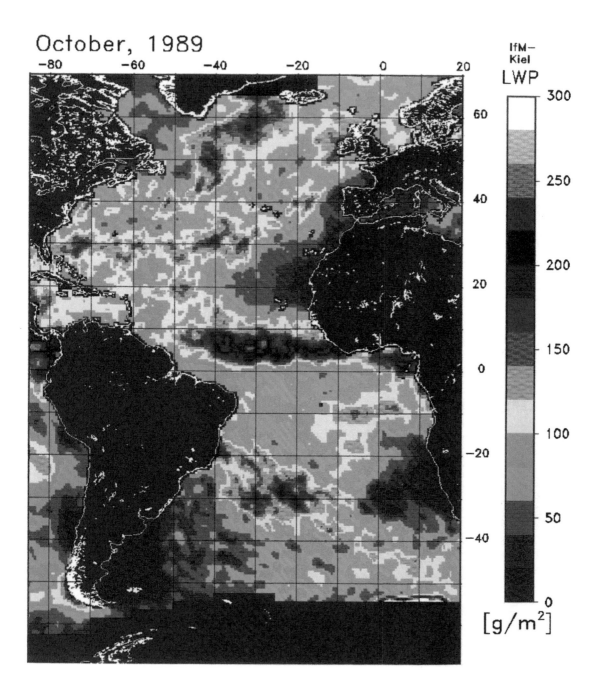

Fig. 11b. Mean monthly liquid water path of clouds over the Atlantic Ocean: October 1989.

Adv. Space Res. Vol. 18, No. 7, pp. (7)17–(7)20, 1996

 Pergamon

0273–1177(95)00283–9

DERIVATION OF PRECIPITABLE WATER FROM METEOR AND NOAA INFRARED WINDOW MEASUREMENTS

A. B. Uspensky and G. I. Scherbina

Hydrometeorological Centre of Russia, 123242 Moscow, Russia

ABSTRACT

This presentation describes the procedures for inferring the precipitable water content (PW) from cloud free infrared radiance observations in the atmospheric window 10.5 - 12.5 μm provided by the METEOR - 3 instrument (single channel) and by the NOAA radiometer AVHRR (two split-window channels). Two types of algorithms for PW retrieval over sea from both kinds of satellite data have been proposed. The first algorithm utilizes the parametric model of satellite measurements and best linear estimator; the second one is based on statistical regression. Verification of remote sensed PW against collocated analysis data gives RMS errors in the range 0.55 - 0.70 cm (0.54 - 0.62 cm) for single - channel (two - channel) methods. The problem of PW retrieval over land with varying surface temperatures and emissivities is under discussion.

INTRODUCTION

A knowledge of water vapour profiles in the atmosphere or total precipitable water (PW) content is required for many applications, notably for assimilation in numerical weather prediction schemes, climate studies. An efficient way to estimate and to monitor global and regional distributions of PW is to use spaceborne observations from meteorological satellites in the thermal infrared (IR) or microwave bands. It was shown by numerous authors that the PW content over sea surface can be inferred from cloud - free IR radiance observations in atmospheric window band 10.5 - 12.5 μm. Such observations are provided by IR radiometers onboard polar-orbiting (METEOR-3, NOAA) and geostationary (METEOSAT, GOES, GMS) satellites. The sensor onboard METEOR-3 (which is called KLIMAT) measures the radiances in channel 10.5 - 12.5 μm with a nominal resolution about 3 km in subsatellite point (now the resolution is ~10 km). The AVHRR and HIRS/2 instruments onboard NOAA have correspondingly two split - window channels in the same band and a channel centered at approximately 11 μm. The geostationary satellite GOMS which is planned to be launched in nearest future has sensor with single IR channel in 10 - 13 μm band.

The purpose of this work is to examine the methods of estimating the PW from cloud free IR radiance observations in 10.5 - 12.5 μm band over sea or over land surfaces. First, the description of technique for PW retrieval over sea will be presented. Then the problem of PW estimation over land surface with varying surface temperatures (Ts) and emissivities is discussed. In this case the synergistic use of AVHRR and HIRS/2 IR window radiance measurements is proposed.

ESTIMATION OF PRECIPITABLE WATER OVER SEA

The feasibility of deriving the PW over sea from cloud free IR measurements in 10.5 - 12.5 μm and some retrieval schemes have been considered since the beginning of 1980th (e. g. /1/-/6/).

Our approach to retrieve PW from /3, 4, 6/ being similar to /1/ is based on the parametric model of IR measurements in the 10.5 - 12.5 μm band. This model has the standard form $y=F*x+e$. Here y is obtained from measurements; x is a vector of unknown parameters with the component $x(1)=(PW-PW0)/PW0$, where PW0 is the priori initial guess for PW; F is a matrix approximating the forward radiative transfer operator and e are the errors with known covariance matrix. Starting from first guess solution with known covariance matrix we employ the best linear estimator to obtain the minimum variance solution.

The quantitative measure of information content of y with respect to x(1) or informativity index R(PW) can be defined as the relative decrease of a priori variance $\sigma^2(x(1))$:

$R(x(1)) = (\sigma^2(x(1)) - \sigma^2(\hat{x}(1)))/\sigma^2(x(1))$, where $\hat{x}(1)$ is the estimate of x(1). The results of $R(x(1))=R(PW)$ calculations for 3 atmospheric models and 3 different levels of measurement errors are presented in table 1, see also /3, 6/. The transmittances in parametric model were computed using procedure from /7/.

Table 1. Theoretical values of informativity index R(PW) for a priory choosen $\sigma(x(1))=0.3$, $\sigma(x(2))=\sigma(x(3))=2K$ and different levels of measurement's error (upper line - for single channel scheme)

PW (cm)	$\sigma(e)$		
	0.5 K	0.3 K	0.1 K
1.72	0.07	0.07	0.07
	0.25	0.41	0.52
3.01	0.30	0.32	0.33
	0.40	0.49	0.58
4.37	0.46	0.51	0.52
	0.56	0.60	0.68

The data from table 1 demonstrate the feasibility of retrieving the PW from satellite data. Basing on these considerations two types of algorithms for PW retrieval from single - channel measurements (KLIMAT) or from split-window channel measurements (AVHRR) have been proposed and evaluated /3, 4/. The first (physical-statistical) algorithm utilizes the parametric model and best linear estimator. The alternative statistical approach to the PW estimation is to employ linear regression estimate. The general feature of the described algorithms is the linear relationship between measured brightness temperatures Tr and the PW.

The experiments with NOAA and METEOR-3 data

NOAA data. The performance of developed technique for PW retrievals has been evaluated in trials with real NOAA data, see /4/. The data set comprising 70 collocated cloud free satellite (AVHRR, HIRS/2), radiosonde (RAOB) and ship observations for the area of North - East Atlantic and Mediterranean (January-June, 1982) has been used for examination of developed algorithms. RMS differences between satellite estimates and RAOB data are in the range 0.55 - 0.65 cm (0.54 - 0.62 cm) for single-channel (two-channel) measurement schemes. It's important to note that the sample for the case study corresponds rather dry atmosphere with the PW changing in the range 1.0 - 2.1 cm and PW = 1.4 cm, $\sigma(PW)$ = 0.48 cm. The statistical retrieval algorithm works slightly better than physical one.

METEOR-3 data. The spectral response functions for IR window channels of METEOR-3 (KLIMAT), METEOSAT and GOMS instruments are shown in fig. 1.

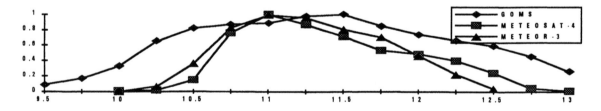

Fig. 1. Spectral response functions of GOMS, METEOSAT-4 and
METEOR-3 (KLIMAT) IR window channel

Taking into account the closeness of response functions for METEOR-3 and METEOSAT instruments the statistical algorithm of PW estimation similar to /5/ has been developed and evaluated. The linear relationship between PW and difference Tr-Ts was found by synthetic regression (see fig. 2). The application of this technique requires the knowledge of Ts and temperature profile at the sounding point.

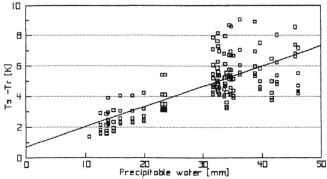

Fig.2. Scatter diagram of (Tr-Ts) versus PW

For the evaluation of this approach the PW retrievals over sea surface (PW. sat) were compared with collocated results of numerical analysis (PW. anal) for the oceans in Northern hemisphere. The retrievals are made for segment areas of 8x8 pixels corresponding to about 80x80 km. RMS differences between PW. sat and PW. anal are in the range 0.62 - 0.70 cm for several dates (orbits) in Aug. 1993 (data from METEOR-3). For more accurate comparison the collection of collocated PW. sat and island radiosonde stations observations as ground truth measurements is required /5/.

RETRIEVAL OF PRECIPITABLE WATER OVER LAND SURFACE

To begin this section it is pertinent to point to the interconnection between PW retrieval technique and split-window method (SWM). The SWM being primarly developed for determination of sea surface temperature has been adapted to PW derivation. The majority of proposed technique for PW retrieval is based on the fact that the difference T4 - T5 between brightness temperatures in the AVHRR split-window channels is sensitive to total water vapour amount along the optical path (or PW). For sea surface a linear relationship is found between T4 - T5 and the PW /2/. This is not so for measurements over land surface. The brightness temperatures difference measured in two AVHRR channels is the function of the surface temperature, atmospheric profiles of humidity and temperature and the channel emissivities. These effects are coupled so that is not possible to separate them with AVHRR data only.

The extension to SWM for retrieving PW over land has been proposed in /8/. It requires a minimum of a priori information but uses the assumption of emissivity invariance. To overcome this restriction the use of combined AVHRR / HIRS data has been proposed. Our approach differs from /9/ and employs matched measurements in the AVHRR channels A4, A5 and in HIRS/2 channel H8 for the estimation of the spectral emissivities.

The suggested method develops the approach /10, 11/ and is based on the concept of the temperature independent spectral index (TISI) constructed for the combinations of measurements in the channels A4, A5 and H8 , see /12/. The TISI's being defined similar to /10, 11/ are practically independent on the surface temperature. It enables to achieve the discrimination between thermal and spectral emissivities effects on recorded IR radiances. Assuming the simple relationship between emissivities for spectral bands of channels A4, A5, H8, the estimates of these quantities can be derived.

While constructing and treating TISI we need a priori information (temperature and humidity profiles in sounding points) to calculate spectral radiances in channels A4, A5, H8 at ground level from satellite measurements. According to /11/ the atmospheric corrections required for calculation of TISI and retrieval of emissivities do not need a very accurate description of the atmosphere and can be performed using climatological data. This fact together with described technique permits to suggest the following procedure of PW estimates refinement over land surface. At first stage we employ the method /12/ with climatological temperature and humidity profiles as input data for estimation of spectral emissivities.

Then at the second stage we use 3 equations constituting single - channel correction method (equation (20) from /13/ for channels A4, A5, H8) or 3 equations similar to /3, 4/ and eliminate unknown parameters (Ts, Ta) to obtain explicit expression for PW.

The efficiency of the proposed technique for emissivities and Ts estimation is proved by the results of experiments with simulated data. Now we proceed with the application of this method to real satellite data. The preliminary results concerning the retrieval of emissivities and LST from experimental data set are promising (/12/).

REFERENCES

1. T. Aoki, T. Inoue, Estimation of precipitable water from the IR channel of the geostationary satellite, *Remote Sens. Envir.*, 12, 219-228 (1982).

2. G. Dalu, Satellite remote sensing of atmosheric water vapour, *Int. J. Remote Sensing, 7*, 1089-1097 (1986).

3. V. M. Sutovsky, A. B. Uspensky, On the use of apriori information for derivation of sea surface temperature based on satellite infrared radiation measurements in 10.5-12.5 μm band, *Issledovanie Zemli iz Kosmosa*, 4, 86-97 (*Soviet Journal of Remote Sensing*) (1985).

4. V. M. Sutovsky, A. B. Uspensky, E. I. Rozanova, Errors in satellite in satellite measured SST and atmospheric water content estimated from outgoing IR (10.5 - 12.5 μm) radiation, *Issledovanie Zemli iz Kosmosa*, 2, 17-23 (*Soviet Journal of Remote Sensing*) (1987).

5. J. Schmetz, L. VandeBerg, Estimation of precipitable water from Meteosat IR window radiances over sea, *Beitr. Phys. Atmosph.*, 5, 93-102 (1991).

6. A. B. Uspensky & V. I. Solovjev, Derivation of sea surface temperature and precipitable water over sea from METEOR and NOAA IR window measurements, *Proc. of the 7th Australasian Rem. Sens. Conf.*, Vol. 3., 223-230 (1994).

7. M. P. Weinreb, M. L. Hill, Calculation of atmospheric radiances and brightness temperatures in IR window-channels of satellite radiometers, *NOAA Tech. Rep. NESS N80*, Washington D. C., 49p (1980).

8. T. J. Kleespiess, L. M. McMillin, Retrieval of precipitable water from observations in the split window over varying surface temperatures, *J. Appl. Meteorol.*, 29, 851-862 (1990).

9. P. Schluessel, Satellite - derived low level atmospheric water vapour content from synergy of AVHRR with HIRS, *Int. J. Remote Sens.*, 10, 705-721 (1989).

10. F. Becker, Z. -L. Li, Temperature independent spectral indices in Thermal IR bands, *Remote Sens. Env.*, vol.32, 17-33 (1990).

11. Z. L. Li & F. Becker, Feasibility of land surface temperature and emissivity determination from AVHRR data, *Rem Sens. Envir.*, 43, 67-85 (1993).

12. A. B. Uspensky, G. I. Scherbina, Land surface temperature and emissivity estimation from combined AVHRR/TOVS data, *Proc. 6th AVHRR data users meeting, Belgirate, Italy, 29. 06 - 02. 07. 1991*, 109-116 (1993).

13. J. A. Sobrino, C. Coll and V. Caselles, Atmospheric correction for LST using NOAA-11 AVHRR channels 4 and 5, *Remote Sens. Environ.*, 38, 19-34 (1991).

Adv. Space Res. Vol. 18, No. 7, pp. (7)21–(7)24, 1996
Copyright © 1995 COSPAR
Printed in Great Britain. All rights reserved
0273–1177/96 $9.50 + 0.00

 Pergamon

0273–1177(95)00284–7

PRECIPITATION INTENSITY ESTIMATION USING AVHRR NOAA DATA

P. Romanov

Hydrometeorological Centre of Russia, 123242 Moscow, Russia

ABSTRACT

The main features of the precipitation diagnosis scheme developed for the AVHRR/NOAA high resolution data are discussed. The classification procedure is based on the maximum likelihood method and includes spectral as well as textural features derived from the satellite measurements. The tuning of the algorithm and the estimation of its effectiveness were performed using ground based synoptic observation data (transmitted via SYNOP reports) synchronous and collocated to the corresponding AVHRR measurements.

INTRODUCTION

The indirect relationship between precipitation and radiation in visible and infrared emerging from the top of the cloud forms a basis for a qualitative method for rainfall estimate from satellite measurements which is widely used in operational meteorology. The known applications of this method mainly relate to geostationary satellite data over tropics and sub-tropics (e.g./1/). The possibility to retrieve information on precipitation over high latitudes with a good spatial and temporal resolution stimulates the efforts to develop corresponding methods making use of the AVHRR NOAA information. Most of the reported automated AVHRR precipitation detection techniques rely on premises similar to geostationary ones (e.g. the higher the albedo and the lower the cloud top temperature, the higher is the precipitation probability or intensity) (see /2,4/). The additional spectral data (for example, channel 3 and 5 measurements) and textural features can provide more information on cloud type and thus are believed to improve the precipitation analysis. It has to be noted that the common problem of all of the developed AVHRR based precipitation estimate techniques is the uncertainty of their accuracy which arises from the lack of validation experiments. Up to now only a few results of qualitative comparison of derived precipitation charts with the predicted amount of precipitation and with synoptic precipitation analysis have been presented /3,5/.

The two year archive of NOAA-11/AVHRR daytime data of maximum horizontal resolution over the Central European part of Russia received in HMC in 1992-1994 and synchronous ground-based conventional observations were used to develop the statistical cloud classification scheme /7/. Precipitation diagnosis is the last stage of this procedure, which is included depending on the result of cloud type and cloud amount estimate made at the previous stages. The decision rule in the precipitation classification procedure as well as in the cloud type classification is the maximum likelihood method. This article presents the brief description of the precipitation estimation method and the results of its validation against ground based precipitation estimates.

METHOD AND DATA

The precipitation intensity estimation is a part of the AVHRR image classification scheme developed for cloud cover analysis. This procedure is included if the cloud amount derived at the previous stage exceed 4 oktas and the cloud is classified as Cb, Ns, Cu, St or Sc. Retrieved cloud characteristics (cloud type, cloud amount) as well as precipitation relate to the 5x5 AVHRR pixel blocks (classification area) which make approximately 9x9 km squares at the ground level in mid-latitudes for the regional Mercator projection with the 1' latitude and 1.5' longitude grid used in this study.

The general classification approach used for precipitation estimation was based on the use of the maximum likelihood method. The scheme included 10 spectral and textural features constructed from AVHRR measurements to separate each 5x5 pixel block into five precipitation categories.
The list of features is given in Table 1, where A_1 and A_2 mean channel 1 and 2 albedo, T_3, T_4 and T_5 mean brightness temperatures in channels 3, 4 and 5. The 700 and 500 mb temperature values (T_{700} and T_{500}) for the centre point of the block, derived from numerical prediction or actual analysis fields and cloud amount N (in percent) obtained by straight forward calculations of cloud-contaminated and cloudless pixels are also used, signs 'min' and 'max' correspond to minimum and maximum values of parameters within the classification area, $< >$ means spatial averaging over classification area and $\triangle$ - is the mean absolute deviation symbol.

TABLE 1   Features used in the precipitation estimate procedure
        (see comments in the text)

1. $< A_1 >$	6. $A_{1max}-A_{1min}$
2. $T_{4min}-T_{700}$	7. $\triangle A_1$
3. $(T_3-T_4)_{max}$	8. $T_{4min}-T_{500}$
4. $< T_4-T_5 >$	9. $< A_2-A_1 >$
5. $\triangle T_4$	10. N

It is quite understandable that the determination of classes included in the classification procedure must take into account the supposed method of validation of obtained results and the source of information used for validation procedure. The present technique is based on the use of qualitative information on precipitation from conventional meteorological stations ('ww' code in SYNOP reports transmitted through GTS (Global Telecommunication System). Thus the precipitation classes were also determined qualitatively: the sets of 'ww' codes describing the present weather situation or observed weather phenomenon were assigned to each of the five categories: 'no precipitation', 'low', 'moderate', 'heavy' and 'risk precipitation' (or 'thunderstorm'). Table 2 presents precipitation categories and corresponding weather 'ww' codes.

TABLE 2   Precipitation categories and corresponding SYNOP 'ww' codes

Category	Codes
1. No precipitation	1-15,19-49
2. Low precipitation	50-59,60,61,66,68,70,71
3. Moderate precipitation	16,62-65,67,69,72-79
4. Heavy precipitation (shower)	80-89
5. Risk precipitation (thunderstorm)	17, 90-99

The mathematical formulation of the statistical classification problem is to determine a certain class i, whose cluster centre in the feature space is the closest to the given feature observation vector **u** or

$$p(u/C_i) \cdot P(C_i) > p(u/C_k) \cdot P(C_k)$$
$$\text{for all } i \neq k$$

where $p(u/C_k)$ - conditional probability function and $P(C_k)$ - class k a priori probability. Usually normal distribution of features vectors within each class is supposed, then the statistics needed to includes features vectors means and matrixes of covariations. To obtain reliable estimates of these parameters as well as estimates of the a priori probability of each class, a certain training period is necessary.

The statistical data on precipitation classes distribution parameters were obtained using a regularly updated archive of collocated synchronous ground based and AVHRR/NOAA-11 daytime observation data. The study area comprised a square with approximately 700 km sides, centred at 37N, 56E. There are over 50 meterological stations transmitting SYNOP reports above the study area, 6 of them are making hourly observations and all the others' reports are available every 3 hours. The maximum time gap between two types of observations subjected to comparison was set to 20 min. Estimates of classification features corresponding to the observation data were made for the 5x5 AVHRR pixel blocks centred at the geographical location of each station. Since the beginning of the experiment, in 1992, more than 3000 synchronous collocated satellite and ground based observations data relating to spring, summer and autumn were accumulated in the archive and used to tune the classification procedure and to estimate its effectiveness.

RESULTS

The routine validation of the developed classification method has started after more than 50 cases of observation of each precipitation class and corresponding AVHRR data were available from the archive. Since then the adjustment of the features distribution parameters was made regularly, using data from each NOAA orbit processed and corresponding synoptic observations. These parameters were recomputed using not less than 50 and not more than 100 latest cases of every class observation. The a priori probabilities of classes 1-5 (Table 2) were set to 0.65, 0.10, 0.08, 0.12 and 0.05 correspondingly according to the frequency of precipitation observations over the study area for the spring-autumn period.

The results of comparison of satellite precipitation rate estimates with precipitation rates derived from SYNOP reports are presented in Figure 1. As it follows from the diagram, the best agreement between satellite and surface data is for 'risk precipitation' and 'no precipitation' categories. Eighty percent of cases with thunderstorm with shower observed were classified as 'risk precipitation' and more than 91% of observations with no rain were classified as 'no precipitation'. Note, that in the latter case, the 9% erroneous classifications comprises approximately 20% of all precipitation observations. Accounting for the misclassifications within other categories, the overall accuracy of separation of precipitating

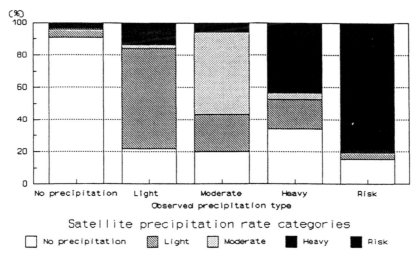

Fig.1 Correspondence between satellite and observed precipitation types

clouds from non-precipitating ones exceeds 85%. The separation of 'light' and 'moderate' precipitation cases is accurate in 55-60% of all cases and 'heavy' precipitation, is classified correctly in 40% of all cases of shower observations. The clear distinction of thunderstorm clouds is understandable, because these clouds differ greatly from all the others types of clouds due to their very low top temperature and high albedo. The good separability of 'light precipitation' class can be explained by relation of this precipitation type to the frontal cloud systems, which are more homogeneous in space and more stable than convective types of cloud. Therefore clusters, related to these cloud types are more stable and more compact. The accuracy of heavy precipitation detection decreases most probably because of the time gap between satellite and ground based observations and due to rapid development of corresponding convective phenomena. All in all, the results of more than 1000 comparisons are accounted in the diagram on Fig.1. The exact fit of satellite precipitation rate estimates to the surface observation data occur in 74% of all cases.

CONCLUSION

Possibilities to derive information on precipitation characteristics using AVHRR/NOAA multispectral data have been investigated. Validation of the satellite soundings against ground truth data revealed the reliability of the developed classification method. The probability of correct separation of raining clouds from non-raining ones when cloud cover exceeds 4 oktas is over 85% and the fit of satellite estimates of precipitation type retrievals to the ground based observations is achieved in 74% of all cases considered. The essential part of all discrepancies is due to the time gap admitted between the two types of observation being compared.

As the rainfall categories used in the classification algorithm are qualitative, the results of classification can not be used in numerical weather prediction or analysis schemes. Thus at the next stage of our study we will try to relate the determined classes to certain precipitation rates on the basis of radar data.

REFERENCES

1. Garand, L., Two automated methods to derive probability of precipitation fields over oceanic areas from satellite imagery, Journ. Appl. Meteorol., 28(9), 913-924 (1989).

2. Inoue, T., An instantaneous delineation of convective rainfall areas using split-window data of NOAA-7 AVHRR, Journ. Met. Soc. Japan, 65(3), 469-481 (1987).

3. Karlson,K.G., Liljas,E., The SMHI model for cloud and precipitation analysis from multispectral data: PROMIS reports, No.10, SMHI, Sweden (1990).

4. Liberti, G.L., Precipitation estimation with AVHRR data: a review, Proc. of the 6th AVHRR data users' meeting, Belgirate, Italy, 29 June -2 July 1993, 31-37 (1993).

5. Pylko, P., Aulamo, H., Using AVHRR data to estimate rainfall in northern latitudes. Proc. of the 5th AVHRR data users' meeting, Tromso, Norway, 25-28 June 1991, 333-337 (1991).

6. Romanov, P.Yu., Cloud parameters and precipitation intensity analysis using AVHRR images classification scheme, Proc. of the 7th Australasian Rem. Sens. Conf., 219-225 (1994).

 Pergamon

Adv. Space Res. Vol. 18, No. 7, pp. (7)25–(7)28, 1996
Copyright © 1995 COSPAR
Printed in Great Britain. All rights reserved
0273–1177/96 $9.50 + 0.00

0273–1177(95)00285–5

PASSIVE REMOTE SENSING OF THE ATMOSPHERIC WATER VAPOUR CONTENT ABOVE LAND SURFACES

B. Bartsch,* S. Bakan** and J. Fischer***

* *Meteorologisches Institut, Universität Hamburg, Bundesstr. 55,
D-20146 Hamburg, Germany*
** *Max-Planck-Institut für Meteorologie, Bundesstr. 55, D-20146 Hamburg,
Germany*
*** *Institut für Weltraumwissenschaften, Freie Universität Berlin,
Fabeckstr. 69, D-14195 Berlin, Germany*

ABSTRACT

The global distribution of the atmospheric water vapour content plays an important role in the weather forecast and climate research. Nowadays there exist various methods dealing with remote sensing of the atmospheric water vapour content. Unfortunately, most of them are restricted to ocean areas, since, in general, the emission of land surfaces is not known well enough.

Therefore, a new method is developed which allows the detection of the atmospheric total water vapour content from aircraft or satellite with the aid of backscattered solar radiation in the near infrared above land surfaces.

The Matrix-Operator-Method has been used to simulate backscattered solar radiances, including various atmospheric profiles of temperature, pressure, water vapour, and aerosols of various types, several sun zenith angles, and different types of land surfaces. From these calculations it can be concluded, that the detection of water vapour content in cloudless atmospheres is possible with an error of < 10 % even for higher aerosol contents.

In addition to the theoretical results first comparisons with aircraft measurements of the backscattered solar radiances are shown. These measurements have been carried out with the aid of OVID (**O**ptical **V**isible and near **I**nfrared **D**etector), a new multichannel array spectrometer, in 1993.

INTRODUCTION

Water vapour is the most important natural atmospheric greenhouse gas, influencing strongly solar and thermal infrared radiativ transfer, and giving rise for clouds, which have strong influence on weather and climate. Unfortunately, water vapour varies considerably in time and space, so that local measurements are not sufficient for the definition of regional and global fields. Therefore, much effort is put into the development of procedures for the global detection of the atmospheric water vapour content from aircraft or satellite.

Nowadays different methods of global water vapour detection exist. Above ocean surfaces, for example, the use of SSM/I (Special Sensor Microwave/Imager) data in the microwave range allows to retrieve water vapour column with mean error of 7 % /1/. Above land surfaces the situation is worse: With the High-Resolution Infrared Sounder (HIRS-2) of TOVS, an operational infrared-microwave sounding system on polar-orbiting platforms, errors in the global detection of 20 % occur /2/, and also the spatial resolution is not suitable for the determination of mesoscale water vapour fields. Also in the near infrared the radiance in the $\rho\sigma\tau$ water vapour absorption band has been proposed with theoretical errors between 7 and 15 % /3/, /4/. But these spectral methods require either the knowledge of the surface temperature or a linear spectral variing surface reflectance. Also these methods are limited to small aerosol contents.

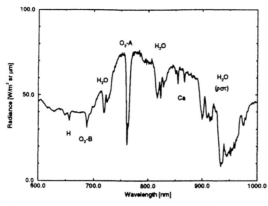

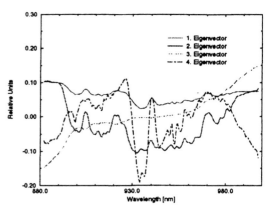

Fig. 1. Airborne measured OVID spectrum of the vertically backscattered radiance above acriculture fields

Fig. 2. First four eigenvectors within the $\rho\sigma\tau$ water vapour absorption band

Therefore we are working towards a water vapour remote sensing algorithm above land surfaces with an error less than 10 % also for higher aerosol contents, and for high spatial resolution.

THEORETICAL INVESTIGATION

Figure 1 shows an example of the vertically backscattered radiance above acriculture fields measured with OVID, our multispectral channel analyzer. The three water vapour absorption bands, both oxygen absorption bands and four Fraunhoferlines (one H and three Ca II) between 600 and 1000 nm can be seen well.

For our simulations the spectral region between 700 and 1000 nm is considered, whereby special attention is given to the $\rho\sigma\tau$ absorption band, since there the broadest range of absorption coefficients occurs, which should result in a good algorithm for low and high water vapour contents. For radiative transfer modelling a Matrix-Operator-Method was used with a spectral resolution of 1.7 nm, which matches the resolution of OVID. The considered 5 standard atmospheres, 13 water vapour profiles with vertical contents from 0.3 - 5.8 g/cm^2, 8 land surfaces with reflectances between 20 and 90 % (whereby lateron also some positive tests with surface reflectances of about 10 % were carried out), 6 aerosol profiles from the WCP-Report 55 /5/ (3 of them with optical depth less than 0.25 and 3 with optical depth of 1.02, 3.07 and 3.32, single scattering albedo reaching from 0.65 up to 0.98) and 6 sun zenith angles resulted in more than 3500 spectra. After adding a noise of 1 % these spectra were subjected to a principal component analysis which is an objective tool to handle big data sets.

Figure 2 shows the first four eigenvectors within the $\rho\sigma\tau$ absorption band with eigenvalues of 236180, 9643, 989, and 96. The first eigenvector is clearly related to the absolut radiance intensity in the window channels. The second shows the spectral feature of the water vapour absorption coefficient in this region and is, therefore, useable for a water vapour algorithm. The third reflects the spectral variations of the underlying surface and the atmospheric aerosols. Finally, since the water vapour absorption coefficients in the $\rho\sigma\tau$ band are highly variable and the eigenvector analysis is a linear tool, while transmission is a nonlinear function of the absorber amount, a fourth eigenvector occurs, which can also be attributed to the water vapour. Probably, this fourth eigenvector can be avoided by using logarithmic radiances within the absorption region as input for the principal component analysis.

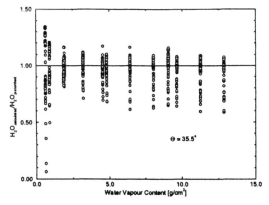

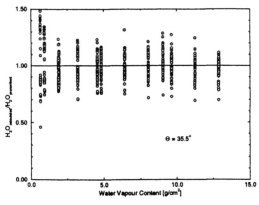

Fig. 3. Ratio of calculated and prescribed water vapour content for different slant water vapour contents without aerosol correction ($\bullet$: τ < 0.25, $\diamond$: $\circ$ > 1.02)

Fig. 4. Same as Fig. 3, but with aerosol correction

If an algorithm is developped only for small aerosols, Figure 3 shows the ratio of calculated and prescribed water vapour content for different values of the latter on the slant path from the sun to the surface and sensor. The rms error for the small aerosol contents (τ < 0.25) is 8 %, while in the three cases with very high optical depths ($1.02 < \tau < 3.32$) the water vapour is underestimated by 40 %.

Using three eigenvectors within the oxygen-A band for aerosol correction the rms error reduces considerably (Figure 4) and can expected to be less than 10 % for a more basic data set.

COMPARISON MEASURED AND CALCULATED SPECTRA

Figure 5 shows measured and simulated normalized radiance spectra with nearly the same atmospheric water vapour content. There exist considerable discrepancies in the spectral behaviour of the spectra within the absorption band. First investigations show, that these discrepancies can be reduced, if the most recent HITRAN data base /6/ for the line by line transmission calculation is used.

SUMMARY AND OUTLOOK

Theoretical simulations show, that with the aid of 4 spectral channels in the $\rho\sigma\tau$ water vapour absorption band and with 3 spectral channels in the O_2-A-absorption band it should be possible to observe the atmospheric water vapour content in cloudless atmospheres from above with an error less than 10 % even for high aerosol concentrations.

In the future new simulations with a more realistic data set of aerosols, a higher vertical resolution and additional water vapour profiles will be carried out. Special attention will be given to the correspondence of measured and simulated spectra. Also the influence of surface elevation with respect to aerosol correction must be investigated.

With the aid of theses simulations the spectral channel number, wavelength location and bandwidth will be optimized. The algorithm will be tested during an aircraft campaign in spring 1995, which will be carried out with our OVID system and a water vapour lidar for comparison.

This investigation will also be used, to define the water vapour channel for MERIS, the Medium Resolution Imaging Spectrometer on board the european satellite Envisat, which will be launched in 1998.

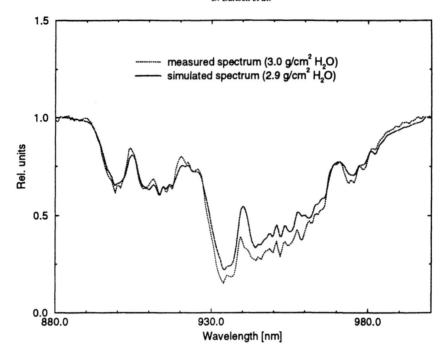

Fig. 5. Comparison of measured and simulated normalized radiance spectra with nearly the same atmospheric water vapour content

REFERENCES

1. P. Schlüssel, W.J. Emery, Atmospheric water vapour over oceans from SSM/I Measurements, *Int. J. Remote Sensing* 11, No. 5, 753-766 (1990).

2. J. Susskind, J. Rosenfield, and D. Reuter, Remote sensing of weather and climate parameters from HIRS2/MSU on TIROS-N., *J. Geophys. Res.* 89, 4677-4697 (1984).

3. B.-C. Gao, A.F.H. Goetz, Ed R. Westwater, J.E. Conel, and R. O. Green, Possible near-IR channels for remote sensing precipitable water vapor from geostationary satellite platforms, *J. Appl. Meteor.* 32, 1791-1801 (1993).

4. R. Frouin, P.-Y. Deschamps, and P. Lecomte, Determination from space of atmospheric total water vapour amounts by differential absorption near 940 nm: Theory and airborne verification, *J. Appl. Meteor.* 29, 448-460 (1990).

5. WCP-55, *Report of the experts meeting on aerosols and their climatic effects*, eds. A. Deepak and H.E. Gerber, Williamsburg, Virginia, 28-30 March 1983.

6. L. S. Rothman et al., The hitran database: editions of 1991 and 1992, *J. Quant. Spectrosc. Radiat. Transfer* 48, No. 5/6, 469-507 (1992).

 Pergamon

Adv. Space Res. Vol. 18, No. 7, pp. (7)29–(7)36, 1996
Copyright © 1995 COSPAR
Printed in Great Britain. All rights reserved
0273–1177/96 $9.50 + 0.00

0273–1177(95)00286–3

RECENT STUDIES ON SATELLITE REMOTE SENSING OF CLOUDS IN JAPAN

T. Hayasaka

Center for Atmospheric and Oceanic Studies, Faculty of Science, Tohoku University, Sendai 980-77, Japan

ABSTRACT

In the latest decade satellite remote sensing of atmospheres, especially of clouds, has been rapidly developed in Japan. A brief review of satellite remote sensing of clouds by Japanese scientists is introduced in this paper. Most of the satellite data used in these studies are obtained by NOAA Advanced Very High Resolution Radiometer (AVHRR). From the infrared split window data which is defined as a brightness temperature difference of AVHRR between channel 4 and channel 5, cloud type classification in the tropical region and identification of clouds over the ice sheet in the Antarctica were successfully carried out. The optical thickness, droplet effective radius, and liquid water path of lower level stratified clouds were obtained from the reflection measurements of solar radiation by visible channel 1 and near infrared channel 3. In addition to AVHRR data analysis, a future perspective of Japanese Earth observation satellite is shortly described.

INTRODUCTION

Moisture in the atmosphere plays quite important roles in climate system of the Earth. Clouds which cover an about half of the Earth's surface are closely related to the radiation budget, precipitation and also evaporation from the land and ocean. Satellite remote sensing is one of the most effective methods for investigating spatial and temporal variations of cloud properties. At present passive remote sensing of clouds from satellite uses a wide range of wavelength from the visible to microwave. Cloud droplets consist of liquid water and/or ice water particles which have large variation of complex refractive index with wavelength /1, 2/. Also a relationship between the wavelength and size of droplet particle must be taken into account in the radiative properties of clouds /3/. That is, reflection, transmission, absorption, and emission properties of cloud change with wavelength, depending on its droplet size distribution and liquid (ice) water content. Therefore this large variation of radiative properties with wavelength enable us to retrieve cloud properties by measuring radiation reflected or emitted from clouds at several wavelengths from the visible to microwave.

This paper presents a brief review of recent studies on satellite remote sensing of clouds in Japan. Japan Meteorological Agency operates a Geostationally Meteorological Satellite (GMS), which gives much contribution to weather forecasting. At present, however, GMS has only two channels, i.e. one in visible region and the other in infrared region. These are not sufficient for investigating cloud properties. The United States meteorological satellite NOAA TIROS series have been used in this field for a long time. The Advanced Very High Resolution Radiometer (AVHRR) on board this satellite has become available in Japan, and some results from AVHRR are shown in this paper.

SPLIT WINDOW DATA ANALYSIS

Classification of clouds

Clouds have large variation with respect to cloud top height, geometrical thickness, optical thickness, and horizontal distribution from which cloud type is classified. The cloud type classification is important to understanding not only clouds' influence on the Earth's climate but also formation and dissipation processes of clouds themselves.

In the infrared window region from 8 to 12 μm, refractive index of water and ice, particularly its imaginary part, varies with wavelength. The variation of refractive index results in the difference in transmissivity and emissivity of clouds. In the case of optically thick clouds, the cloud reaches blackbody so that brightness temperature observed from space does not change with respect to wavelength. On the other hand, in the case of optically thin clouds such as cirrus clouds, the clouds are not blackbody so that radiation from the underlying surface and atmosphere passes though the cloud and a part of radiation thermally emitted from clouds is added to upward radiation above clouds. Therefore the brightness temperature of optically thin clouds changes with wavelength, and these clouds are detectable by means of two channels in the infrared window region.

In 1980's classification of clouds using the AVHRR split window data which is defined as a difference of brightness temperature between 11 μm (ch4) and 12 μm (ch5), was successfully performed by Inoue /4, 5/. For example, tropical clouds were classified by difference of brightness temperature and brightness temperature at 11 μm as shown in Figure 1 /5/. In Figure 1 blue indicates cumulonimbus clouds. Cirrus-type clouds are shown by green, red, and pink, corresponding to optically thick cirrus clouds, moderate cirrus clouds, and optically thin cirrus clouds, respectively.. Yellow area is low-level stratus or stratocumulus. White indicates cumulus

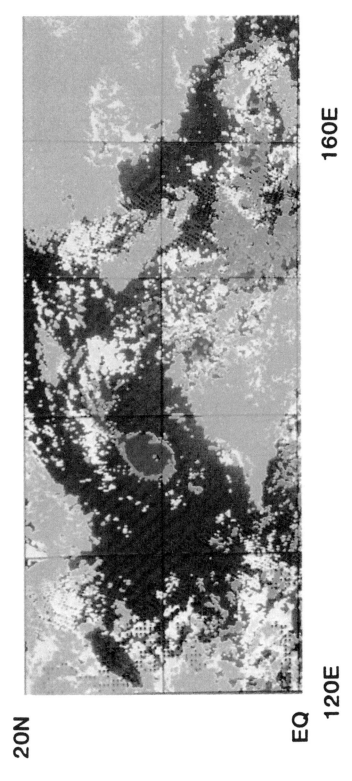

Fig. 1. Cloud type classification map corresponding to Fig. 1. Blue indicates cumulonimbus clouds (B). Cirrus-type clouds are shown by green (I1), red (I2), and pink (I3). Yellow area is low-level stratus or stratocumulus (U). White indicates cumulus cloud field (N). Light blue indicates clear and black indicates no data or coast line.

20N

EQ

120E

160E

T. Hayasaka

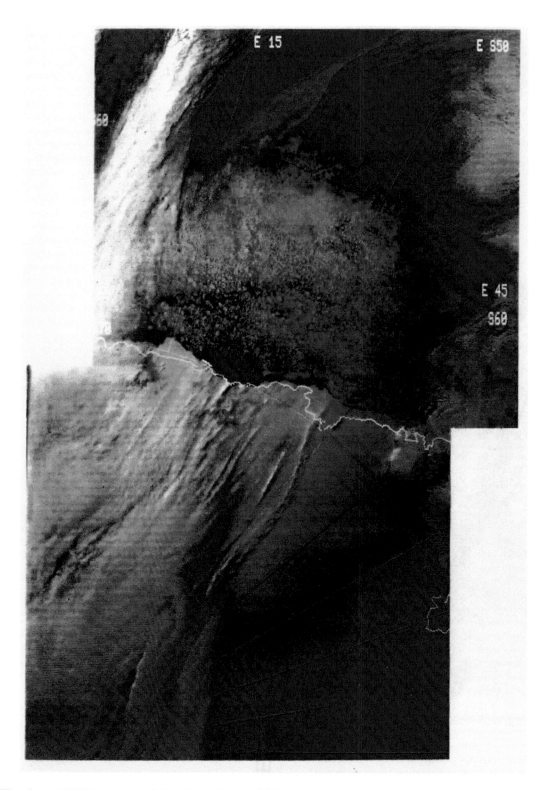

Fig. 2. AVHRR image of clouds (white and blue), Antarctic continent covered by ice sheet (orange), sea ice (orange), and open sea (black seen in the upper part).

cloud field. Light blue indicates clear and black indicates no data or coast line. Various types of clouds are seen around the typhoon, especially cirrus clouds extend in wide area. Semi-transparent clouds are hard to identify only from visible or infrared data. However, the split window technique results in good cloud classification although quantitative retrieval of cloud physical properties such as optical thickness, droplet size or liquid (ice) water path is still insufficient.

Identification of clouds over ice sheet in Antarctica

As an another example of application of the split window technique, an identification of clouds over the ice sheet in the Antarctica were carried out by Yamanouchi /6, 7/. It is difficult to identify the clouds over ice sheet from visible and/or infrared imagery because radiative properties of ice sheet are closely similar to those of clouds, i.e. high reflectivity and low temperature. However, there are small differences in the infrared split window properties between ice sheet and clouds. There is little difference in the split window of ice sheet while clouds have brightness temperature difference as long as optical thickness is not so large. It is, therefore, possible to discriminate cloud pixels from underlying ice sheet pixels.

Yamanouchi also showed from theoretical calculation that thermal emission part contained in 3.7 μm (ch3 of AVHRR) data is quite effective to discriminating clouds from ice sheet /6/. This is attributable to a principle that emissivity at 3.7 μm (ch3) does not approach unity even in the large liquid or ice water path. For the large optical thickness, a sign of brightness temperature difference changes. At present, however, its successful application has not been performed because of low data quality of ch3. In ch3 of AVHRR reflected solar radiation is also included in the daytime data and is useful to discriminate clouds from ice sheet.

Figure 2 shows an example of cloud detection in the Antarctica from AVHRR ch1, ch2, and a difference between ch3 and ch4 /8/. Clouds shown by white color, over ice sheet shown by orange are clearly seen both in the marine and continent regions. Since AVHRR has rather high spatial resolution (1.1km just under the orbit), a group of ice floating in the sea is clearly shown.

ESTIMATION OF OPTICAL THICKNESS, DROPLET SIZE, AND LIQUID WATER PATH

Quantitative estimation as well as qualitative understanding of various properties of clouds is indispensable for studying climate of the Earth. The optical thickness, cloud liquid water path, and droplet effective radius can be retrieved from the radiance of solar radiation in visible and near infrared spectral regions /9/. The reflected radiance in the visible is governed almost only by optical thickness of clouds because clouds hardly absorb radiation in this spectral region. In the near infrared, on the other hand, cloud droplets slightly absorb radiation so that the reflected radiance depends not only on the optical thickness but also on the droplet size distribution. Since the optical thickness of clouds consists of droplet size distribution and liquid water path, the liquid water path is also obtained from the analysis of AVHRR ch1 and ch3. Since the ch3 radiance includes thermal emission as well as reflected solar radiation, it is removed from the data by using ch4 infrared data.

In this retrieval, multiple scattering has to be taken into account in the radiance calculation. In the late 1980's Nakajima developed an accurate and rapid transfer code for calculating the intensity field /10/. Therefore ch1 and ch3 of AVHRR have been used for the quantitative remote sensing of clouds. Using these two channels, the optical thickness, liquid water path and droplet effective radius of winter stratocumulus clouds in the Western North-Pacific were retrieved /11/. Figure 3 shows an example of the relationship between reflected radiance and cloud parameters such as optical thickness at 0.64 μm and droplet effective radius. As shown in Figure 4, cloud optical thickness and effective radius are well determined, i.e. a relationship between them is almost orthogonal except for the case of small optical thickness.

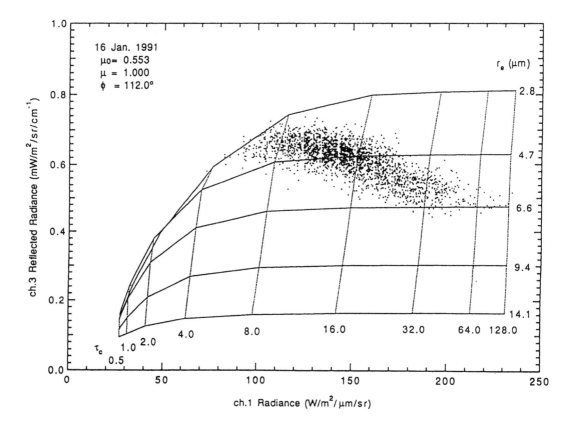

Fig. 3. Relationship between the reflected radiance at ch1 and ch3 of AVHRR on January 16, 1991 for various values of the cloud optical thickness and droplet effective radius.

The results thus obtained from AVHRR data were compared with in situ measurements by airborne 37 GHz microwave radiometer and Particle Measuring System Inc. Model FSSP-100 cloud droplet spectrometer. The microwave radiometer provides liquid water path from the brightness temperature difference measurements between cloudy region and clear region. From FSSP-100 measurements cloud droplet size distribution is obtained in a size range from 2 to 47 μm with an interval of 3 μm in diameter. The comparison shows that, both in respect of the effective radius and liquid water path, the retrieved values and in situ measurements are consistent.

SUMMARY AND FUTURE PERSPECTIVE

In this paper some of activities in Japan on satellite remote sensing of clouds were briefly introduced. Although the topics were limited to AVHRR data, from which classification of cloud types, identification of clouds over ice sheet, and quantitative estimation of cloud microphysical properties were carried out, satellite data other than AVHRR are also being used in the field of atmospheric sciences. Among them, analysis of microwave data such as Special Sensor for Microwave Imager (SSM/I) is being developed in Japan. Microwave remote sensing is quite effective to estimating moisture content in the atmosphere and surface in spite of lower spatial resolution.

In addition to utilization of state-oft-the-art satellite data, Japanese Earth observation satellites are being developed. Advanced Earth Observing Satellite (ADEOS) will be launched by National Space Development Agency (NASDA) in 1996. After ADEOS, Tropical Rainfall Measuring Mission (TRMM) is planned for launch in 1997 under the cooperation with NASA, USA. TRMM will directly measure rainfall rate with radar as well as with microwave imager from space. For observing atmospheric moisture, ADEOS-II is now being planned, which has Advanced Microwave Scanning Radiometer (AMSR) and Global Imager (GLI; it consists of 36 channels from visible to infrared). Therefore contributions to atmospheric moisture estimation not only from data analysis but also from the viewpoint of hardware are expected in Japan.

Acknowledgments. The author is grateful to T. Inoue, T. Yamanouchi, M. Kuji, and T. Nakajima for their cooperation in the presentation of this paper.

REFERENCES

1. G. M. Hale and M. R. Querry, Optical constants of water in the 200-nm to 200-μm wavelength region, Appl. Opt., 12, 555-563 (1973).
2. S. G. Warren, Optical constants of ice from the ultraviolet to the microwave, Appl. Opt., 23, 1206-1225 (1984).

3. H. C. van de Hulst, Light scattering by small particles, Dover, New York, 470pp. (1981).

4. T. Inoue, A cloud type classification with NOAA-7 split window measurements, J. Geophys. Res., 92, 3991-4000 (1987).

5. T. Inoue, Features of clouds over the tropical Pacific during northern hemispheric winter, J. Meteor. Soc. Japan, 67, 621-637 (1989).

6. T. Yamanouchi, K. Suzuki, and S. Kawaguchi, Detection of clouds in Antarctica from infrared multispectral data of AVHRR, J. Meteor. Soc. Japan, 65, 949-962 (1987).

7. T. Yamanouchi and S. Kawaguchi, Cloud distribution in the Antarctic from AVHRR data and radiation measurements at the surface, Int. J. Remote Sensing, 13, 111-127 (1992).

8. T. Yamanouchi and K. Seko, Antarctic from NOAA satellites (clouds, ice and snow), National Institute for Polar Research, Japan, 91pp. (1992).

9. T. Nakajima and M. D. King, Determination of the optical thickness and effective particle radius of clouds from reflected solar radiation measurements. Part I: Theory, J. Atmos. Sci., 47, 1878-1893 (1990).

10. T. Nakajima and M. Tanaka, Algorithms for radiative intensity calculations in moderately thick atmospheres using a truncation approximation, J. Quant. Spectrosc. Radiat. Transfer, 40, 51-69 (1988).

11. T. Hayasaka, M. Kuji, T. Nakajima, and M. Tanaka, Satellite remote sensing and air-truth validation of cloud liquid water path and droplet effective radius, Proceedings of the 8th Conference on Atmospheric Radiation, 23-28 January 1994, Nashville, Tennessee, American Meteorological Society, 421-422 (1994).

Adv. Space Res. Vol. 18, No. 7, pp. (7)37–(7)40, 1996
1995 COSPAR
Printed in Great Britain
0273–1177/96 $9.50 + 0.00

 Pergamon

0273–1177(95)00287–1

ANGULAR EFFECT IN AVHRR'S SPLIT-WINDOW SEA SURFACE TEMPERATURE AND ATMOSPHERIC MOISTURE OVER THE ATLANTIC OCEAN

A. Ignatov and G. Gutman

Satellite Research Laboratory, NOAA/NESDIS, Washington, DC 20233, U.S.A.

ABSTRACT

The AVHRR on board NOAA satellites observes the underlying surface at different zenith angles within $\Theta \approx \pm 68°$ around nadir. The algorithms for retrieval of sea surface temperature (SST) t_s and the column water vapor content W from the angle-dependent brightness temperatures (BT) $t_i(\Theta)$ (i=4 and 5 for AVHRR Channels 4 and 5 with central wavelengths λ=10.8 and 12 µm, respectively) should overtly account for Θ to provide a result invariant of the variable observation geometry. Recently, a statistical method based on empirical angular functions (EAF) was proposed to assess the angular effect in both the original AVHRR BT's and retrieved SST and to test angular methods of SST retrieval /1/. The EAF's were employed to analyze the BT's in AVHRR Channels 3 and 4 and dual-window SST over tropical Atlantic in Jun 1987 and Dec 1988 from NOAA-10 and -11, respectively /1/. Here, we extend it to check the split-window SST and W, and the angular method of SST retrieval /2,3/, over the same target in tropical and an additional target in North Atlantic from NOAA-9 in Jul 1986.

CONCEPT OF EMPIRICAL ANGULAR FUNCTIONS (EAF)

BT in the i-th Channel $t_i(s;m)$ under cloud-free conditions depends upon the state of the ocean-atmosphere system (vector s) and observation geometry ($m=\sec\Theta$ - relative air mass of the atmosphere). If the ocean and atmosphere were constant (s=const) then BT would depend only on geometrical factor $t_i(s;m) \equiv t_i(m)$. We define that dependence as angular function. The simplified radiative transfer equation for aerosol-free atmosphere and non-reflective sea surface gives /4,5/:

$$t_i(\Theta) \approx t_s - (t_s - \bar{t}_a)\cdot W\cdot K_i\cdot m \qquad (1)$$

where $\bar{t}_a$ is the effective temperature of the atmosphere; K_i-absorption coefficient. According to (1), for a given vector $s=\{t_s, \bar{t}_a, W\}$=const, $t_i(m)$ may be approximated linearly with respect to m. The derivation of the empirical angular function - EAF - from the top-of-the-atmosphere satellite measurements is, however, not trivial since it requires observations of a target with a constant SST through a constant atmosphere at many viewing angles Θ_k, k=1,..,K. The ATSR on board ERS1 allows viewing any target on the Earth surface at two angles $\Theta_1 \approx 0°$ and $\Theta_2 \approx 55°$ from two consecutive points of the orbit. The statistical procedure proposed in /1/ allows one to estimate the EAF's in the full range of AVHRR viewing angles. It is based on a histogram analysis of satellite data collected over a large quasi-uniform ocean region $S \sim (10°)^2$ which is stable enough during the time $\tau \sim 1$ month to allow multiple observations of S at different viewing angles. The BT's histograms are constructed using all satellite measurements over (S,τ) but separately for different angle bins $\Delta_k \equiv (m_k, m_k+\Delta m)$. The warm quasi-gaussian peak is associated with clear-sky views, and its mode $\bar{t}(\Delta_k)$ -- the mathematical expectation of the BT over (S,τ,Δ_k) domain -- corresponds to the mean ocean-atmospheric condition $\bar{s}$. Estimation of the modes of histograms for different angle bins Δ_k (k=1,..,K) provides K points in the sought EAF, since all the mean values $\bar{t}(\Delta_k)$ are associated with the mean ocean-atmosphere state $\bar{s}$, which is the same for different angle bins.

Application of the EAF. According to (1), the SST can be estimated either by a formal extrapolation of $t_i(m)$ to m=0 /2,3/ or as a linear combination of BT's in two channels. (1) underlies the NOAA's algorithm MCSST /6/ and suggested in /7/ algorithm for W retrieval

$$t_s = \alpha_1\cdot t_4(m) + \alpha_2\cdot t_5(m) + \alpha_o; \qquad W = A\cdot\{t_4(m)-t_5(m)\}/m \qquad (2,3)$$

where $A=[(K_{\lambda5}-K_{\lambda4})\cdot(t_s-\bar{t}_4)]^{-1}=1.96$ g·cm^{-2}·(°C)$^{-1}$ is assumed constant. The EAF's $t_4(m)$ and $t_5(m)$ enable one to check whether the linear approximation (1) holds and provides consistent intercepts in two channels, and the algorithms (2,3) result in angle-invariant SST and W.

DATA

In the present study, we use daily images from the NOAA's Global Vegetation Index (GVI) data set /8/. GVI is a subsample of Global Area Coverage (GAC) data mapped into a Plate Carree projection with a (0.15°)² lat/lon resolution. Since Apr 1985, it consists of global maps of the AVHRR counts in the Channels 1, 2, 4, and 5, as well as the associated scan and solar zenith angles. The GVI primary mission has been global vegetation studies (although the data over oceans was mapped also). As a result, the temperature was perceived to be of auxiliary role: the 10-bit GAC counts in Channels 4 and 5 are converted to the BT's T_4 and T_5 and further truncated to an 8-bit format /8/. This truncation results in a loss of accuracy: the GVI BT's are digitized with a step of 0.5°C for t≥-30°C. The non-linearity correction was done using the tabulated data of /9/. The absolute radiometric accuracy of resulting BT's is estimated to be no worse than 1°C. The described satellite BT data are far from ideal for EAF estimation because of low spatial resolution (an order of magnitude cruder as compared to GAC data and two orders cruder as compared to LAC) and crude 0.5°C-digitalization. We employed that data because of their convenient structure, and to provide a simpler consistent case for comparison with similar land studies which are underway. The statistical nature of the procedure implies that the use of high resolution data could improve the quantitative results of the present paper.

31 daily images from NOAA-9 for Jul 1986 were processed over two regions: S_1 in tropical Atlantic (10-20°N, 40-50°W -the same as used in /1/) and S_2 in N.Atlantic (50-62°N, 10-30°W). To estimate the homogeneity and stability over those regions, we used the SST-climatology /10/ which provides the multi-annual monthly means t_c and standard deviations σ_c within (1°)² grid boxes. That climatology contains 121 (1°)²-boxes over S_1 and 271 over S_2, respectively. The mean $\bar{t}_c$ over S_1 and S_2 are 26.3 and 12.5°C, respectively. The value of σ_c, which characterizes the year-to-year variability in SST, is ≈1°C over tropical and ≈1.2°C over N.Atlantic. According to /10/, ~34% of the SST values can be expected between $\bar{t}_c$ and $\bar{t}_c+1\cdot\sigma_c$; ~14% between $\bar{t}_c+1\cdot\sigma_c$ and $\bar{t}_c+2\cdot\sigma_c$; and ~2% between $\bar{t}_c+2\cdot\sigma_c$ and $\bar{t}_c+3\cdot\sigma_c$, and a mirror image is found on the negative side of the mean. One should bear that in mind when comparing the results of SST retrieval with climatological values in the next Section. The spatial heterogeneity of the SST field within S_1 and S_2 is quantitatively described by standard deviations of t_c-values within S_1 and S_2 boxes of 0.5°C and 1.4°C, respectively.

The total AVHRR swath was subdivided into 18 angular bins (±9, symmetrically with respect to nadir) with an equidistant step of $\Delta m=0.2$. Crude filters (thresholding reflectances in Channel 1 and spatial uniformity tests in Channels 1 and 4) were used to exclude most obvious clouds, the objective being to clean up the peaks of the histograms to allow more reliable estimate of the position of the mode. An example of normalized smoothed histograms of the BT's over S_1 and S_2 for one angle bin is shown in Fig.1. Over tropical Atlantic, the shape is close to the Gaussian, similarly to described in /1/. Over the N.Atlantic, the two-peak shape indicates the presence of two clusters on the underlying surface. This feature is traced in almost every angle bin. We have fit this distribution with (bi)normal function

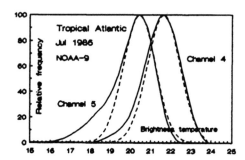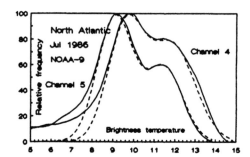

Fig.1. Histograms of BT's in Channels 4 and 5 in the near-nadir angular bin over tropical and N.Atlantic. Dashed lines show the results of their fitting by (bi)normal distributions.

TABLE 1. Statistics of the EAF's linear fit.

	Chan	Regression	R	σ, °C
Tropical Atlantic	4	24.0 - 2.0·m	0.99	0.19
	5	22.9 - 2.2·m	0.97	0.32
North Atlantic	4	11.5 - 1.8·m	0.93	0.39
	5	11.7 - 2.5·m	0.94	0.50

and have chosen the more statistically significant colder cluster assuming it represents typical conditions for this region, whereas the warm peak most probably results from Gulf Stream intrusion.

RESULTS

The results of the $\bar{t}(\Delta_k)$ derivation are given in Fig.2. The histograms in two angular bins over tropical Atlantic were extremely wide and have been excluded from the analysis. The statistics of the linear fits are given in Table 1. The results over the second target are less accurate because of pronounced non-uniformity of underlying surface and persistent cloudiness. In tropics, intercepts in Channels 4 and 5 differ both from each other and the climatic SST (26.3°C) significantly. Over the N.Atlantic, extrapolation to m=0 provides SST more compatible with the climatic norm (12.5°C) in both Channels.

The operational in Jul 1986 split-window MCSST algorithm used the following formula: $t_s=3.4317·t_4-2.5062·t_5+1.58$ /11/. Substituting the t_4 and t_5 values from Table 1 for m=1 and 2 into this equation yields: $t_s=25.2$ and 23.8°C for tropics, and 11.8 and 11.9°C for the N.Atlantic, respectively. Thus, in high latitudes, MCSST for this particular case performs satisfactorily. In tropics, however, it underestimates the SST, with the descrepancy increasing with the viewing angle. The latter result agrees with the conclusion drawn in /1/ for the dual-window MCSST. The EAF's in split-window channels in tropical Atlantic are almost parallel, similarly to the dual-window ones, and no angle-independent coefficients in linear retrieval algorithm (2) can provide here angle-invariant SST. Use of the angle dependent coefficients was proposed in /12/. The model coefficients α_i were tabulated as a function of m for two samples of atmospheric situations representative for N.Atlantic and tropics. We applied those coefficients to the respective EAFs from Table 1, for m=1 and 2. The results are 24.9 and 24.5°C for tropics, and 10.8 and 11.4°C for N.Atlantic. The formulations /12/ treat angular dependence better than MCSST, yielding, however, the SST lower than both MCSST and the climatic norm.

Similarly to SST algorithms, one can check the algorithm for water vapor retrieval. Substituting EAF's from Table 1 into formula (3) for m=1 and 2 yields: W=2.5 and 1.5 g·cm^{-2} in tropics, and 1.0 and 1.2 g·cm^{-2} over N.Atlantic. W have no W-cimatology in cloudless conditions to validate the above values.

The angular effect in the retrieved SST and W is more pronounced in tropical regions. That may imply that the errors come most probably from simplified atmosphere treatment rather than the black surface approximation, used in deriving (1), because in tropical regions the warm downwelling radiance compensates for most of the deficit in emissivity, and additionally the surface signal is strongly attenuated by atmosphere, further depressing the effect of non-unit emissivity.

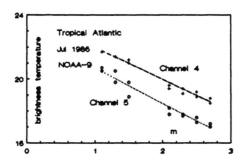
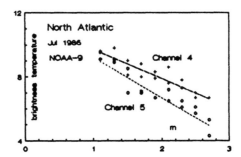

Fig.2. EAF's of BT's in Channels 4 and 5 and their linear fit over tropical and N.Atlantic.

CONCLUSION

The EAF's allow one to check consistency of different angular methods for SST retrieval in different spectral regions. The EAF's estimated over tropical Atlantic show that the results of extrapolation to zero air mass underestimate SST as compared to climatic norm in the region up to 2.5°C in Channel 4 and 3.5°C in Channel 5. For the N.Atlantic, the results of extrapolation in both channels are close to climatic SST within 1°C. The EAF's allow one to check whether the algorithms for the SST and *W* retrieval using cross-scanning radiometer data provide angle-invariant retrievals. Both the MCSST /6/ and water vapor /7/ retrieval algorithms fail to treat the angular effect in the extremal moist conditions. This result is in close agreement with the one obtained in /1/ over the same target in tropical Atlantic using two independent data sets. The accuracy of the EAF's derivation has to be improved. It can be done by using high-resolution satellite data of brightness temperatures with more precision. We plan to carry out similar analyses over selected land surfaces to assess the angular effect in the derived land surface temperature and water vapor over land. Theoretical and model analysis of the algorithms for the sea surface temperature and integral water vapor retrievals should go along with experimental studies.

Aknowledgements. Thanks go to Dr.G.Ohring (SRL) for critical review of the manuscript. D.Sullivan (Research and Development Corporation) helped us in organizing the AVHRR time series over the two oceanic targets, and P.Clemente-Colon (SRL) provided the SST-climatology. This work was done when A.I. held National Research Council Associateship at SRL, on leave from the Marine Hydrophysics Institute, Sevastopol, Crimea, Ukraine.

REFERENCES

1. A.Ignatov and I.Dergileva, Angular effect in dual-window AVHRR brightness temperatures over oceans. *Int.J.Rem.Sens.*, in press.

2. A.Gorodetsky, Estimation of surface temperature by the angular scanning method. *Issledovanie Zemli iz Kosmosa*, #2, 36-44 (1981, in Russian).

3. A.Chedin, N.Scott and A.Berroir, A single-channel double-viewing angle method for SST determination from coincident METEOSAT and TIROS-N radiometric measurements. *J.Appl.Meteor.*, 21(4), 613-618 (1982).

4. C.Prabhakara, G.Dalu and V.Kunde, Estimation of sea surface temperature from remote sensing in the 11- to 13-μm window region. *J.Geophys.Res.*, 79(33), 5039-5044 (1974).

5. L.McMillin, Estimation of sea surface temperatures from two infrared window measurements with different absorption. *J.Geophys.Res.*, 80(36), 5113-5117 (1975).

6. P.McClain, W.Pichel and C.Walton, Comparative performance of AVHRR-based multichannel sea surface temperature. *J.Geophys.Res.*, 90(C5), 11587-11601 (1985).

7. G.Dalu, Satellite remote sensing of atmospheric water vapor. *Int.J.Rem.Sens.*, 7(9), 1089-1097 (1986).

8. K.Kidwell, Global Vegetation Index User's Guide, U.S. Department of Commerce, National Oceanic and Atmospheric Administration, 49pp (1990).

9. M.Weinreb, G.Hamilton, S.Brown and R.Koszor, Nonlinearity corrections in calibration of Advanced Very High Resolution Radiometer infrared channels. *J.Geophys.Res.*, 95(C5), 7381-7388 (1990).

10. U.S.Navy Marine Climatic Atlas of the World, 1992. Naval Oceanography Command Detachment, Asheville, N.C., CD-ROM Ver. 1.0, March 1992.

11. I.Barton, Satellite-Derived Sea Surface Temperatures -- A comparison between operational, theoretical, and experimental results. *J.Appl.Meteorol.*, 31(5), 433-442 (1992).

12. D.Llewellyn-Jones, P.Minnett, R.Saunders and A.Zavody, Satellite multichannel infrared measurements of sea surface temperature of the N.E.Atlantic Ocean using AVHRR/2. *Quart.J.Roy.Meteor.Soc.*, 110, 613-631 (1984).

Pergamon

Adv. Space Res. Vol. 18, No. 7, pp. (7)41–(7)46, 1996
Copyright © 1995 COSPAR
Printed in Great Britain. All rights reserved
0273-1177/96 $9.50 + 0.00

0273–1177(95)00289–8

THE ESTIMATION OF MOISTURE OVER EGYPT FROM METEOSAT SATELLITE OBSERVATIONS

M. A. Mosalam Shaltout,* A. H. Hassan* and T. N. El-Hosary**

* National Research Institute of Astronomy and Geophysics, Helwan, Cairo, Egypt
** Meteorological Authority, Abbasiya, Cairo, Egypt

ABSTRACT

In Egypt their are 77 ground stations for recording the meteorological elements, but very rare number of these stations cover the desert, while the desert covers 97% of the total land of Egypt. For that, estimation of meteorological data over the desert from Meteosat observations is very interested for the national programs for the desert development.

We selected seven ground stations, each one of them represents a specific climatic region in Egypt, for a correlation analysis between the water vapour absorption observed by Meteosat in the spectral band (5.7-7.1 µm) and evaporation and relative humidity measured at these seven ground stations for the period (1985-1986). The correlation coefficients are good for the evaporation, and acceptable for the relative humidity.

Also, a correlation analysis of cloudiness observed by Meteosat in the thermal spectral band (10.5-12.5 µm) with surface infra-red measurements at Cairo and Aswan for the period (1990-1992) was performed. The correlation coefficients are good for Cairo, but a weak for Aswan.

The infra-red measured from ground by Eppley Precision Infrared Radiometers for the period (1990-1992) have excellent correlation with the air temperature at Cairo and Aswan. Also, there is a good correlation with water vapour at Cairo, but a weak correlation for Aswan. Similar we found a good correlation with cloudiness observed from ground at Cairo, and weak for the observations at Aswan.

Models and empirical relations for estimating the moisture over Egypt from Meteosat observations are deduced and tested.

INTRODUCTION

Most of Egypt land (97%) is desert area, and there are national programs for developing it. Remote sensing is very important techniques for Geological, Geophysical ,Geographical, and Meteorological studies for the development of the desert. Remote sensing techniques are the only means to interpolate between the very few direct measurements in the atmosphere and at ground. Their data, if assimilated appropriately into numerical analysis schemes, will provide also as a basis for more accurate estimates of the global water budget and its regional components, thus allow for better observations of future climate variations.

The aim of the present study is to develop and test new improved models for using Meteosat observations to estimate moisture in the atmosphere of Egypt.

DATA

1. Satellite data :

The Meteosat satellite is stationed in a geostationary orbit at nearly 36.000 km above the equator and the Greenwich meridian (O°N, O°E). The principle payload of the satellite is a multi spectral radiometer /1/. This provides the basic data of the Meteosat system. The three channel radiometer includes :

- Two identical adjacent visible channels in the 0.4-1.1 µm spectral band.
- A thermal infra-red (window) channel in the 10.5-12.5 µm spectral band.
- An infra-red (water vapour) channel in the water vapour absorption band (5.7-7.1 µm), which can be operated on place of one of the two visible channels.

TABLE 1 Ground Meteorological stations

No.	Station	Climatic region	Coordinates of Station		
			Lat. N	long. E	Elev. (m)
1	Matruh	Mediterranean	31° 20`	27° 13`	28.3
2	El Arish	Sinai	31° 16`	33° 45`	15.0
3	Tahrir	Nile Delta	30° 39`	30° 42`	15.6
4	Cairo	Cairo Area	30° 05`	31° 17`	34.4
5	Asyout	Middle Egypt	27° 03`	31° 01`	234.7
6	Khargha	Western Desert	25° 27`	30° 32`	77.8
7	Aswan	Upper Egypt	23° 58`	32° 47`	200.0

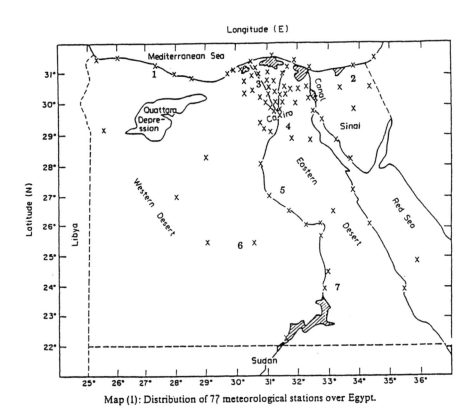

Map (1): Distribution of 77 meteorological stations over Egypt.

The sets of images in any one half hour period are the 2.5 km resolution visible, the 5 km infra-red (11 µm) and the 5 km water-vapour (6 µm).

Every day at L.M.T. 11^h, the Meteorological Authority of Egypt at Cairo, receives images from the satellite Meteosat to analyze cloudiness over Egypt and the surrounding countries. One image is in the visible spectral band (0.4-1.1 µm), and the other in the thermal infra-red window (10.5-12.5 µm). Additionally, a third image in the water vapour absorption band (5.7-7.1 µm) is received. The brightness of each pixel as seen by the satellite is interpreted as an index of atmospheric opacity, which in turn enables the calculation of surface insulation using models or empirical relations /2,3/. We classified the brightness within five bins (very dark areas are zero, while very bright areas are four), and measured the daily brightness of the cloudiness in the spectral band (10.5-12.5 µm) for Cairo and Aswan for the period (1990-1992). Also, the daily brightness in the water vapour spectral band (5.7-7.1 µm) are measured for seven sites in Egypt (as in table (1) and map (1)) for the period (1985-1986).

2. Ground data :

We selected seven ground stations from 77 ground meteorological stations over Egypt, each station from the seven stations represent a specific climatic region in Egypt (as in table (1)). The relative humidity in percent, and the evaporation in millimeters are measured at the seven stations for more than 25 years /4/.

For the both stations Cairo and Aswan we used additional ground data, such as the ambient air temperature in degrees celsius (°C), sky cover in oktas, and solar infra-red radiation measured by Eppley Precision Infra-red Radiometers.

ANALYSIS AND DISCUSSION

Considerable difficulties were encountered in obtaining images of water vapour absorption band (5.7-7.1 µm) per day on a regular basis from Meteosat during the period (1985-1986). Some difficulties arose from the operation of the receiving facilities at the Meteorological Authority in Cairo, and others from poor quality and low contrast of the images. The monthly average of the water vapour absorption in the spectral band (5.7-7.1 µm) are determined for the seven stations of table (1).

A regression analysis is carried out for correlations between relative humidity and evaporation measured at ground and water vapour absorption observed by Meteosat :

1st degree $$Y_i = a_1 + b_1 X$$

2nd degree $$Y_i = a_2 + b_2 X + C_2 X^2$$

3rd degree $$Y_i = a_3 + b_3 X + C_3 X^2 + d_3 X^3$$

where   i = 1,2. Y_1 is the monthly average of evaporation in millimeters (EV), and Y_2 is the monthly average of relative humidity (R.H).
X is the monthly average of water vapour absorption observed by Meteosat (W.V).

Also, a simple linear regression between the Infrared measured from ground with cloudiness measured in thermal Infrared band by Meteosat, cloudiness measured from ground, water vapour and air temperature measured at ground were performed :

$$Y = a + b X \quad \text{for daily and monthly averages.}$$

Table (2) shows the result of the regression analysis between the monthly average of EV and R.H with W.V for the period (1985-1986) over seven sites of Egypt.

From table (2) we can notice that :

1. The correlation between the evaporation at the ground and the water vapour absorption observed by Meteosat are higher than the correlation with relative humidity for the seven sites over Egypt.
2. The correlation increases with increasing degree of the regression.
3. The correlation at Cairo and Tahrir is relatively lower than for the other five sites. It is fair for EV and weak for R.H. For the othert five sites it is very good for EV and good for R.H.

The constants of the empirical relations (Models) in table (2) are tested by calculating the standard error of the estimations for the evaporation and relative humidity using the water vapour absorption observed by Meteosat. It is found the standard errors of estimations are acceptable and the models are reliable.

Table (3) shows the correlation coefficients between the infra-red solar radiation measured at ground by Eppley Precision Infrared Radiometers and the cloudiness observed by Meteosat in the thermal spectral

M. A. Mosalam Shaltout *et al.*

TABLE 2· Regression between the monthly average of EV and R.H with W.V for (1985-1986) over seven sites of Egypt.

Cities	a	b	c	d	r	SE	Y
Matruh 1	8.714	-0.632			0.859	0.402	EV
2	8.54	0.0614	-.273		0.895	0.369	EV
3	8.512	0.3502	-0542	0.0626	0.896	0.390	EV
Matruh 1	70.013	-2.26			0.642	2.886	R.H
2	71.9	-9.81	2.976		0.862	2.009	R.H
3	72.72	-18.05	10.63	-1.786	0.892	1.904	R.H
El-Arish 1	5.074	-0.433			0.801	0.344	EV
2	4.896	0.540	-0.4114		0.912	0.248	EV
3	4.92	0.182	-0.051	-0.09	0.914	0.261	EV
El-Arish 1	73.155	-2.163			0.749	2.034	R.H
2	73.388	-3.432	.536		0.756	2.116	R.H
3	73.636	-7.142	4.277	-0.935	0.764	2.216	R.H
Tahrir 1	8.871	-1.45			0.659	1.712	EV
2	8.443	0.702	-0.918		0.707	1.697	EV
3	8.732	-3.487	3.324	-1.064	0.726	1.751	EV
Tahrir 1	62.0	1.464			0.2306	6.387	R.H
2	63.99	-8.532	4.261		0.473	6.1	R.H
3	63.096	4.433	-8.867	3.293	0.5046	6.336	R.H
Cairo 1	9.51	-1.502			0.625	1.87	EV
2	8.958	1.229	-1.197		0.695	1.816	EV
3	9.289	-3.517	3.811	-1.306	0.716	1.869	EV
Cairo 1	53.23	0.749			0.1065	6.972	R.H
2	55.55	-10.75	5.042		0.449	6.604	R.H
3	54.54	3.735	-10.25	3.988	0.484	6.857	R.H
Assyout 1	19.8	-5.114			0.772	3.625	EV
2	19.1	-2.6	-1.13		0.777	3.779	EV
3	21.877	-23.04	22.5	-6.935	0.823	3.618	EV
Assyout 1	32.124	5.535			0.485	8.585	R.H
2	36.4	-9.872	6.925		0.589	8.36	R.H
3	30.586	32.97	-42.58	14.533	0.675	8.1	R.H
Khargha 1	23.278	-6.654			0.839	3.5	EV
2	23.46	-7.26	0.272		0.84	3.686	EV
3	24.59	-15.333	10.341	-3.137	0.845	3.853	EV
Khargha 1	22.46	8.17			0.709	6.6	R.H
2	24.33	2.09	2.732		0.721	6.836	R.H
3	22.66	14.05	-12.189	4.65	0.727	7.184	R.H
Aswan 1	26.98	-6.796			0.861	3.197	EV
2	27.173	-7.303	-2.8		0.861	3.369	EV
3	27.99	-11.64	5.167	-1.458	0.863	3.554	EV
Aswan 1	12.944	8.315			0.728	6.238	R.H
2	16.191	-0.129	3.47		0.748	6.368	R.H
3	13.716	13.045	-11.59	4.43	0.756	6.657	R.H

r   means correlation coefficient

S.E. means standard error of estimation

TABLE 3 The correlation coefficients between Infrared Solar
radiation measured from ground with other
Meteorological Elements

Site	Element of correlation	Daily average				Monthly average			
		1990	1991	1992	Total	1990	1991	1992	Total
Cairo	I.R + T	0.955	0928	0.92	0.93	0.986	0.95	0.97	0.97
	I.R + W	0.69	0.59	0.72	0.67	0.845	0.75	0.887	0.83
	I.R + SC	-0.71	-0.587	-0.67	-0.631	-0.929	-0.827	-0.89	-0.87
	I.R + GC	-0.69	-0.576	-0.67	-0.638	-0.94	-0.767	-0.906	-0.86
Aswan	I.R + T	0.967		0.95	0.95	0.989		0.986	0.99
	I.R + W	0.39		0.288	0.317	0.630		0.45	0.52
	I.R + SC	-0.31		-0.178	-0.22	-0.669		-0.32	-0.45
	I.R + GC	-0.26		-0.16	-0.20	-0.66		-0.30	-0.39

TABLE 4 Regression Coefficients of simple linear equation for I.R
measured from ground with T, W.V, SC, and GC
for Cairo and Aswan for (1990-1992).

City	Elements of regression	1990		1991		1992		Total	
		a	b	a	b	a	b	a	b
		Regression For Monthly Data							
Cairo	I.R + T	-20.07	0.088	-16.96	0.077	-21.89	0.094	-20.024	0.088
	I.R +W.V	4.63	0.087	4.87	0.069	4.27	0.106	4.554	0.090
	I.R + SC	6.44	-0.38	6.25	-0.312	6.45	-0.42	6.371	-0.073
	I.R + GC	6.78	-0.335	6.25	-0.310	6.621	-0.539	6.632	-0.488
Aswan	I.R + T	-10.77	0.05			-12.30	0.055	-11.75	0.053
	I.R + W.V	3.029	0.149			3.215	0.123	3.145	0.134
	I.R + SC	4.53	-0.339			4.43	-0.25	4.486	-0.300
	I.R + GC	4.60	-0.683			4.43	-0.24	4.47	-0.326
		Regression For Daily Data							
Cairo	I.R + T	-8.2	0.04	-8.2	0.02	-8.2	0.02	-8.123	0.035
	I.R + W.V	2.03	0.03	2.1	0.02	1.9	0.03	2.004	0.03
	I.R + SC	2.6	-0.08	2.55	-0.07	2.55	-0.07	2.57	0.077
	I.R + GC	2.6	-0.10	2.59	-0.07	2.6	-0.10	2.6	-0.095
Aswan	I.R + T	-4.4	0.02			-5.4	0.02	-5.084	0.023
	I.R + W.V	1.6	0.02			1.6	0.02	1.57	0.025
	I.R + SC	1.8	-0.03			1.8	-0.02	1.813	-0.030
	I.R + GC	1.8	-0.03			1.8	-0.02	1.813	-0.034

T   is the temperature of ambient air.

W   is the water vapour measured from ground

SC   is the cloudiness observed by Meteosat in the spectral band
        (10.5-12.5 µm).

GC   is the cloudiness observed from the ground station.

Total   means the total period.

band (10.5-12.5 μm). The table also shows the correlation between the infrared measured of the ground and the temperature of the air, the water vapour measured at the ground, and the cloudiness observed from the ground. Table (4) shows the coefficients of regression a and b.

From Table (3) we can notice that :

1. The correlation of infrared measurements at the ground with cloudiness observed by Meteosat in thermal infrared band is very good for all the years (1990-1992) for Cairo. It is good for Aswan in 1990 and weak in 1992.
2. Generally the correlation for the monthly averages are higher than the correlation for the daily averages.
3. The best correlation for the daily averages are between the infrared measurements at the surface with the temperature of the ambient air, then with the water vapour measured of the ground, then with the cloudiness measured at the ground and space.

This high level of correlation helped to find empirical models of low standard error of estimation to estimate the moisture in the atmosphere of Egypt on the basis of simple linear regressions as in table (4).

CONCLUSION

In the study empirical formulas (Models) for the estimation of the moisture in the atmosphere within seven climatic regions of Egypt were discussed.

From a correlation analysis between the water vapour absorption observed by Meteosat in the spectral band (5.7 - 7.1 μm) and evaporation and relative humidity at ground, it is found the correlation is good for the evaporation, and acceptable for the relative humidity.

Also, the correlation between infrared radiation measured of the ground with cloudiness observed from Meteosat in the thermal infra-red band (10.5-12.5 μm) was performed for the period (1990-1992). It is found the correlation is good for Cairo and weak for Aswan.

The obtained empirical models can be used to estimate the moisture over the desert of Egypt, where the ground stations are very rare in these remote areas. Studing the seasonal and regional variation of moisture over the desert is very interested object for the national programs for the desert development.

REFERENCES

1. J. Morgan, Introduction to the Meteosat, European Space Operation Centre ESOC, Darm Stadt, Federal Republic of Germany (1978).
2. M.A. Mosalam Shaltout, and A.H. Hassan; Solar Energy Distribution Over Egypt Using Cloudiness From Meteosat Photos, Solar Energy Vol. 45, No. 6, pp. 345-351 (1990).
3. W. Mösr and E. Raschke; Incident solar radiation over Europe estimated from Meteosat data, J. Climate Applied to Meteorology 23, pp. 166-170 (1984).
4. Climatological Normals for the Arab Republic of Egypt up to 1975; Meteorological Authority, Cairo, Egypt (1979).

Adv. Space Res. Vol. 18, No. 7, pp. (7)47–(7)51, 1996

Pergamon

0273–1177(95)00288–X

SURFACE MOISTURE REMOTE SENSING WITH SSM/I DATA

J. R. Givri and S. H. Souffez

Groupe Physique-Environnement, Université de Rennes, Campus de Beaulieu, 35042 Rennes, France

ABSTRACT

The total energy budget at the terrestrial surface is explained in terms of heat transfer and evaporation, including wind energy, in a tentative to relate them to the difference of temperature between air and ground at the surface. Both parameters are modeled to assess the surface moisture concentration and evaporation fluxes, taking into account the SSM/I channels temperatures and emissivities at various frequencies in the microwave region : 18, 22, 37, 85 GHz. Data under consideration is supposed to be cloud decontaminated.

INTRODUCTION

A previous study /1/ showed that the 85 GHz channel of the Special Sensor Microwave/Imager (SSM/I) can be used more easily than other channels for surface temperature T_S assessment, the analysis of SSM/I channels at 19, 22 and 37 GHz providing too much light values for T_S. The differences between the T_S values under consideration being related to the atmospheric water content, the present study is intented to emphasis the modeling of the heat transfer at the surface before its radiative transfer up through atmosphere to space, particularly as regard the 37 GHz frequency which is suspected to be related to surface evaporation (EVP) assessment. Similarly, the 85 GHz is suspected to be related to surface wind assessment too. The observation of the previously mentioned difference between the various T_S values infers the idea that these differences could be related as well to the water delienation on surface, which can be assessed by surface emissivity variables, obtained by direct radiative transfer equation retrieval, as to many atmospheric disturbance effects. Thus, the difference between T_S values can also be suspected to be related to the difference between air and surface temperatures T_a and T_S, respectively, which could at its turn provide the excepted EVP assessment /2/.

METHOD

SSM/I data are provided by the Defense Meteorological Satellite Program (DMSP)-F8 satellite. The SSM/I instrument is connically at 53.1° incident angle scanning total power superheterodyne receivers radiometer type. All measurements are obtained with dual polarization except for the 22 GHz channel. The addition of an 85 GHz channel makes the SSM/I unique to previous instruments like the Scanning Multichannel Microwave Radiometer (SMMR) or the Electrically Scanning Microwave Radiometer (ESMR). The SSM/I swath on Earth is 1,394 km, the highest resolution (14 km) ever flown for a passive microwave radiometer, this provides the highest sensitivity for measuring scattering surface or mediums and precipitation, sometimes helpful for atmospheric effects assessment in addition with others channels. In clear sky conditions, taking into account the RAYLEIGH-JEANS assumption, the radiative transfer equation leads to :

$$T_v^{V,H} = \varepsilon_v^{V,H}\,\tau_v(p_s)T_s + \int_{ps}^{0} T(p)d\tau_v(p) + \left(1-\varepsilon_v^{V,H}\right)\tau_v(p_s)\int_0^{Ps} T(p)d\tau_v(p) \qquad (1)$$

where $T_v^{V,H}$ designates the microwave dual polarized (V = vertical, H = horizontal) brillancy temperature obtained by combination of surface and atmospheric emittance in addition with its ortion reflected by the surface up through atmosphere to space. The surface emissivity, surface pressure atmospheric transmission and averaged thermal profile being designated : $\varepsilon_v^{V,H}$, ps, $\tau_v(p_s)$ and T_a, respectively, the relationship (1) leads to, /1/ :

$$T_v^{V,H} = \varepsilon_v^{V,H} \tau_v(p_s)T_s + \varepsilon_v^{V,H} T_a (1 - \tau_v(p_s)) \tag{2}$$

The previous relationship leads directely to the difference between both emissivities as follows, /1/ :

$$\Delta T_v = T_v^V - T_v^H = \tau_v T_s \left(\varepsilon_v^V - \varepsilon_v^H\right) = \tau_v T_s \Delta\varepsilon_v \tag{3}$$

$(1 - \tau_v(1p_s))$ being considered as a minor contribution to emissivities difference determination and taking into account : $(1 - \tau_v(1 - \varepsilon_v)) \approx \varepsilon_v$.

RESULTS AND DISCUSSION

a) The exact assessment of radiating fluxes at the surface
They require to take into account accurately water or heat cycles at the surface. Unfortunately, the complexity of the global problem remains so high that it cannot be successfully solved or modeled without the slightest assumption. Generally, the terrestrial ecosystem can be considered as remaining in a sort of macroscopic thermodynamical equilibrium where the solar net radiation Rn(t) balances permanently upthrough atmosphere evaporation, convection and conduction fluxes to each others, designated Φ_L , Φ_C and Φ_K , respectively. It gives :

$$R_n(t) + \Phi_L(t) + \Phi_C(t) + \Phi_K(t) = 0 \tag{4}$$

The problem consists now to estimate correctly the daily fluxes, designated $\overline{\Phi}_L, \overline{\Phi}_C, \overline{\Phi}_K$ and usually used from their instantaneous values $\Phi_L(t)$, $\Phi_C(t)$ and $\Phi_K(t)$. For example, the density of evaporation fluxe Φ_L can be expressed as follows :

$$\Phi_L = \varphi_e(W).\left(e_s - e_a\right) \tag{5}$$

where $\varphi_e(w)$ designates the heat exchange coefficient between the dry air and the water vapor, related to the surface wind W. e_s and e_a designate the saturated water vapor pressures at the surface and at some level in the atmospheric boundary layer, respectively. Both of these water vapor pressures remain in proportion with humidity concentrations at the surface and within the atmospheric boundary layer, and designated Ks and Ka, respectively. The previous formula shows that $\overline{\Phi}_L$ can be easily extracted from the difference $\overline{e}_s - \overline{e}_a$, which can be related at its turn to the daily values of Ts and Ta respectively. Otherwise, the density of convection fluxe Φ_C within the atmospheric boundary layer can be expressed as follows :

$$\Phi_C = \varphi_C(W).\left(T_a - T_s\right) \tag{6}$$

where $\varphi_C(W)$ designates the convection exchange coefficient, in the atmospheric boundary layer, which can be expressed as follows :

$$\varphi_C(W) = \rho C_p \left(\mu + \mu' W^\alpha\right) \tag{7}$$

In the previous relationship, ρ designates the air density, C_p its heat capability at constant pressure and (μ, μ') a couple of two constant coefficients which expresses the thermal losses at the surface and due to the wind drying action on soil or sea within the boundary layer. μ and μ' are experimentaly determined at an elevation of 2 m above the surface, (which can be considered as the mean rugosity height for various types of soils covered or not by vegetation). The coefficient α is determined at its turn by the type of vegetation : 0.5 (forests) $\leq \alpha \leq 2$. (sea), /3/. The SSM/I pixel size being under consideration, (18 km approximately in scan B mode and 50 km in scan A mode), the vegetation heigth action can be considered as neglectable and then $\alpha \approx 1$. Otherwise, $\varphi_e(W)$ and $\varphi_c(W)$ are not independant, they remain closely connected to each others by the air moisture psychometric coefficient designated (= 0.65 hPa/K), as follows :

$$\varphi_c(W) = \gamma \, \varphi_e(W) \tag{8}$$

The study of the daily variation of Φ_C shows that Φ_C remains included between 100 W/m^2 and 500 W/m^2 and then is graded similarly to R$_n$. Within the process expressed in (1), only R$_n$ and Φ_C are dominating terms at the difference of Φ_L and Φ_K which graduation is estimated to be included between 1 W/m^2 and 100 W/m^2, respectively, /3/. Therefore, only daily values can be reasonnably assessed for comparison with satellite data. Furthermore, the measurement of the daily value of Φ_K shows that it is equal to zero. The evaporation density Φ_L can be related to the difference between water concentrations at the surface Ks and within the atmospheric boundary layer Ka, as follows :

$$\Phi_L = L. \varphi_e(W).\left(K_a - K_s \right) \tag{9}$$

where Φ_L designates the water latent heat of vaporization (2x10^6 J/kg). Both expressions of Φ_L can be combined taking into account relationships (7), (8), (9), it leads to :

$$T_a - T_S = \lambda \left(K_a - K_s \right) \tag{10}$$

The parameter $\lambda = \dfrac{ML\gamma}{RC_p}$ being under consideration, where M designates the molar mass of water vapor and R the universal perfect gas constant, that parameter γ can be assimilated to the temperature of 2.5 x 10^4 K. The proportion of water vapor concentration at the surface being designated $\dfrac{\Delta K}{K} = \dfrac{K_a - K_S}{K_a}$, it leads to :

$$\frac{\Delta K}{K} = \frac{T_a - T_S}{\lambda} \tag{11}$$

b) The EVP assessment

It can be shown, /4/, that the evaporation fluxe density Φ_L can be expressed as follows :

$$\Phi_L = \frac{ML}{RT} \left(\frac{e_a - e_s}{r_a + r_c} \right) \tag{12}$$

when r_a and r_c designate the aerodynamic resistance and the stomatal resistance of the vegetation cover, respectively. r_a is related to $\varphi_c(W)$ by : $r_a = \dfrac{\rho C_p}{\varphi_c(W)}$. Both r_a and r_c are similar to the inverse of a speed (in sec/m). With the wind W estimated at 2 m heigth over a grass field, r_a remains included between 10 and 1450 sec/m. Otherwise, the minimum value for re is 60 sec/m. Then, relationship in (12) leads to :

$$\Phi_L = L. \varphi_c(W) \frac{K_a - K_s (T_S / T_a)}{\rho \, C_p - r_c \, \varphi_c(W)} \tag{13}$$

The stomatal resistance over land surface is given by :

$$r_c = \frac{\rho C_p K_S \left(T_a - T_S\right)}{\varphi_C (W) T_a \left(K_a - K_S\right)} = r_a \left(\frac{\Delta T}{T}\right) surf. / \left(\frac{\Delta K}{K}\right) surf. \tag{14}$$

Taking into account the relationship in (11), the real evaporation $\overline{\Phi}_L$ can be expressed as follows :

$$\overline{\Phi}_L = \beta \, \overline{K}_a \, T_a^2 \tag{15}$$

with :

$$\beta = \frac{R}{M\gamma(\gamma + \gamma_R)} \, , \qquad \gamma_R = \frac{4\varepsilon\sigma T_a^3}{\varphi_e (W)} \tag{16}$$

where ε and σ designate the thermal infrared emissivity of the surface and the BOLZTMAN constant respectively. The previous relationship (15) is very important for three main reasons. It shows that Φ_L varies in proportion to the water vapor concentration in the atmosphere, and thereby Φ_L varies in proportion to the saturated water pressure on the vegetation cover. At last, it explains the influence of the infrared radiation in Φ_L assessment. But the more important is to be shown that Φ_L varies in proportion with the square of the air temperature T_a. More particularly, it shows that any increase of air temperature of 10 % produces an increase of 20 % of the evaporation. Over desert areas like Sahel, it could be the explanation of the role of high air temperature. Over septentrional countries like Europe, the air moisture is included between 0.35 kg/m^3 and 0.70 kg/m^3, infering a variation of about 70 % between these extreme values. Ta varies at its turn between 250 K and 310 K, the variation being of about 20 % only. That explains why Ta - TS remains particularly uneasy to retrieve from satellite data, even with sensors designed for the previous assessments, /5/, /7/.

The present solution can be given by the relationship (7) which provide the wind W related to Φ_L, taking into account the relationship (8). As mentioned in /6/, the wind W can be retrieved from 85 GHz measurements. It leads to :

$$W \approx 3 \left(\frac{T_{85}^V - T_{85}^H}{T_{85}^H}\right) + 1.5 \tag{17}$$

$\varphi_c(W)$ can hence be calculated, taking into account the relationship (14). Then, K_S is given by the following expression :

$$\overline{K}_S = \frac{r_c \, \varphi'_c(W) \, T_a \, (\Delta K) surf.}{(\Delta T) surf.} \tag{18}$$

The previous relationship provides the moisture of 0,37 kg/m^3, nearby the surface (60 % of water vapor in the atmosphere). The satellite measurement providing T_a or T_S nearby the surface, r_a is also retrieved by the same process. Finally, the formula (18) leads to the expected moisture K_a or K_S.

CONCLUSION

The previous method can be essentially estimated as a tentative to assess the surface parameters with microwave data only. The analysis of the problem shows that the critical parameter remains the difference T_a - T_S. Other topics show that the humidity coefficients K_a and K_S, related to T_a and T_S can be easily infered from T_a and T_S.

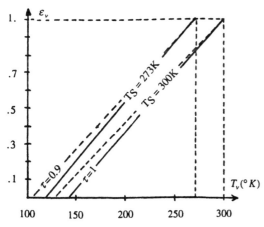

Fig 1 : The surface emissivity versus
the brillancy temperature T_v for various
surface temperature T_S and transmittivity

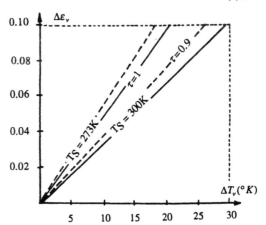

Fig 2 : The emissivity difference
versus the temperature difference
various surface temperature T_S and
transmittivity .

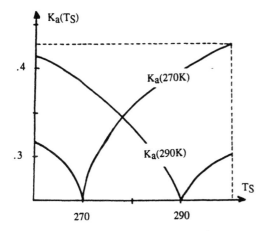

Fig 3 : Showing the influence of
surface temperature for K_a assessments.

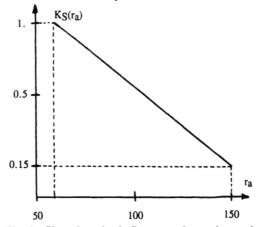

Fig 4 : Showing the influence of aerodynamic
resistance ra for K_S assessment.

REFERENCES

1. J.R. Givri, S.H. Souffez, Surface moisture assessment with microwave and infrared data,
IEEE-IGARSS' 94, Los Angeles, USA, (1992).
2. C. Vorosmarty, A. Grace, B. Moore, A strategy to study regional hydrology and terrestrial
ecosystem processes using satellite remote sensing, ground based data and computer modeling,
Acta Astronautica, Vol. 25, #12, pp. 785-792, (1990).
3. A. Perrier, A. Vidal, Analysis of a simplified relation for estimation daily crapotranspiration
for satellite thermal IR data, Int. Jal. of Rem. Sens., Vol 10, #8, pp. 1327-1337, (1989).
4. A. Vidal, Y. Kerr, J.P. Lagouarde, B. Seguin, Teledetection et bilan hydrique utilisateur
combinée d'un modèle agrométéorologique et les données thermiques du satellite NOAA-
AVHRR, Agricultural and Forest Meteorology, 39, p. 155-175, (1987).
5. K.B. Katzaros, G.W. Petty , On the remote sensing of the maine atmosphere with the Special
Sensor Microwave/Imager (to appear).
6. C. Ottle, D. Vidal-Madjar, G. Gira, Remote sensing application to hydrological modilling, (to
appear in Jal of Hydrology and Geology).
7. M. Raffy, F. Becker, An inverse problem occuring in remote sensing in the thermal infrared
bands and its solutions, Jal. Geophy. Res., 90, #D3, pp. 5809-6819, (1985).

Pergamon

Adv. Space Res. Vol. 18, No. 7, pp. (7)53–(7)62, 1996
Copyright © 1995 COSPAR
Printed in Great Britain. All rights reserved
0273-1177/96 $9.50 + 0.00

0273–1177(95)00290–1

COMPARISON OF RAINGAUGE ANALYSES, SATELLITE-BASED PRECIPITATION ESTIMATES AND FORECAST MODEL RESULTS

B. Rudolf, H. Hauschild, W. Rüth and U. Schneider

Global Precipitation Climatology Centre, Deutscher Wetterdienst, Postfach 10 04 65, D-63004 Offenbach, Germany

ABSTRACT

Observed precipitation data are required for the verification of weather prediction and climate models. Information on precipitation is available from different observation techniques, on global scale from networks of raingauge stations and from satellite remotely sensed data. Both data sources include certain deficiencies. Raingauge measurements are nearly not available over the oceans. They are point measurements, which are locally influenced, and are contaminated by systematic measuring errors. However, these conventional measurements give the best estimate of the true precipitation amount, if the network density is sufficient. Several climatologies based on conventional observations have been published. Satellite data cover land and oceans and provide area-means, but the precipitation estimates are based on indirect approachs and need to be calibrated and validated. This paper will give some intercomparisons, which have been carried out in the framework of the World Climate Research Programme's Global Precipitation Climatology Project.

THE GLOBAL PRECIPITATION CLIMATOLOGY PROJECT

With regard to global climate and its change, precipitation is especially important as its spatial distribution identifies the regions of maximum latent heat release, which is the major driving force of the atmospheric circulation. Ground-based raingauge measured precipitation data are available from most areas of the continents, but generally not from the ocean-surface, which contributes the major part to the earth's surface heat and water fluxes. Satellite-based precipitation estimates are needed to fill this gap, even if these estimates are derived by the use of radiation models or other indirect approachs.

The Global Precipitation Climatology Project (GPCP) was established by the World Climate Research Programme to provide global data sets of area-averaged and time-integrated precipitation based on all suitable observation techniques at least for the period 1986-1995 on a 2.5° grid latitude by longitude /19/. The GPCP is integrated in the Global Energy and Water Cycle Experiment (GEWEX).

The project has been realized by the colaboration of several groups and is organized into satellite- and land-based components: For estimation of precipitation from satellite observations, infrared images are utilized by the Geostationary Satellite Precipitation Data Centre (GSPDC at Climate Analysis Center, Washington D.C.), and microwave data by the Polar Satellite Precipitation Data Centre (PSPDC at Goddard Space Flight Center, Greenbelt MD). The GPCP also contains a component for calibration and validation of the satellite based precipitation algorithms with several activities, as the Surface Reference Data Centre (SRDC at NOAA, NCDC, Asheville NC), an Algorithm Intercomparison Program /2/ and the development of new measuring methods.

The Global Precipitation Climatology Centre (GPCC, operated by the Deutscher Wetterdienst, National Meteorological Service of Germany) is the main land-surface component. It also takes over the satellite-based estimates and numerical weather prediction model results and merges the individual results to a complete global product /15/. The terrestrial gridded analyses are combined with the satellite-based estimates preliminary using a simple blending scheme /7/. To obtain complete global data sets remaining gaps are filled with numerical weather prediction results accumulated from daily forecasts of the European Centre for Medium-Range Weather Forecasts (ECMWF, Reading).

Up to now, the GPCC has prepared terrestrial raingauge-only analyses as well as completed global data sets for the period January 1987 to December 1988 /6/ /7/. Besides this global product, the GPCP gridded datasets based on the individual sources are available, too, as listed in Table 1.

The products are distributed to users via the World Data Centres for Meteorology (Asheville, NC, and Obninsk, Russia). The global merged and the terrestrial raingauge-only products are also directly provided by the GPCC on special request. The products are preliminary since only a limited number of raingauge stations, about 6,700 worldwide, was included.

TABLE 1   The different GPCP data sets (monthly data)

Data set name (grid size)	Data set description	Period available
"GPCP global merged data set, V01" (2.5° lat/lon)	Preliminary global data sets merged by the GPCC	Jan. 1987-Dec. 1988
"GPCP tropical satellite-IR data set" (2.5° lat/lon)	Satellite-based data set between 40°N-40°S derived from IR (and OLR) images evaluated at the GSPDC	Jan. 1986-Mar.1994
"GPCP oceanic satellite-microwave data set" (5° lat/lon)	Satellite-based data set over ocean areas between 60°N-60°S derived from microwave images evaluated at the PSPDC	July 1987-Mar. 1994 (except Dec. 1987)
"GPCC terrestrial rain-gauge analysis" (2.5° and 1° lat/lon)	Raingauge analysis, including the number of stations used on the grid (totally 6700 stations), carried out by the GPCC	Jan. 1987-Dec.1988

COMPARISON OF THE RESULTS FROM DIFFERENT DATA SOURCES

At first, the GPCC preliminary results are compared to other data on a global and terrestrial basis, including products provided by Hulme /10/, Jaeger /11/, Leemans and Cramer /13/, Legates /14/ and Schemm et al. /16/. Additionally to the observational results, the ECMWF weather predictions (spectral model T106, daily 12-36h forecasts accumulated to monthly totals) are included in this study.

The analyses of the GPCC are based on mothly totals of quality-controlled raingauge data obtained from SYNOP and CLIMAT messages exchanged via the World Weather Watch Global Telecommunication System complemented by data from several global and national data collections. At present, data from worldwide about 6,700 stations are processed. The spatial interpolation on grid points and calculation of gridded areal-means is done using an objective analysis method /21/.

The terrestrial part of the climatologies is generally based on evaluated raingauge data. The results of the various authors are derived from different data collections and by different interpolation schemes, since the goals of the authors differed. Hulme, for example, compiled time series of monthly estimates in order to investigate climate trends. His results on 5° grid-cells are based on temporally most-homogeneous input data. Under this condition, about 4,000 stations could be used for the period from 1950 onwards, and a larger number of grid-cells are without results /10/. Legates aspired to the best-possible spatial resolution (0.5° grids), and included long-term monthly means on dissimilar and partly unknown periods, but of ca. 25,000 stations /14/). The IIASA data set prepared by Leemans and Cramer is based on data of the defined period 1931-1960 with 6,000 stations. Schemm analysed the terrestrial precipitation month by month for the period 1979 to 1989 which was based on about 1,400 station data published periodically in reports of the World Monthly Surface Station Climatology. The "Schemm, MSU" dataset is a combination of the MSU satellite based estimates over oceanic and the Schemm analysis over land areas. All authors used objective interpolation methods exept of Jaeger, who utilized continental and national climate atlases containing subjectively constructed isohyetal maps.

Legates performed two versions, the non-corrected "nc" one interpolated from the station data as reported, and the corrected product "c" using the same stations, but data corrected with regard to the systematic errors resulting from evaporation losses and drift due to wind. The assessment of this error is highly important if the conventional data are used for validation of satellite-based estimates or climate model results.

Only Jaeger and Legates provide complete global data without using satallite data. Jaeger's oceanic results were assembled from a global mean yearly precipitation map and monthly precipitation-frequency maps of the U.S. Marine Climatic Atlas. Legates included estimates from ship weather observations.

An overall intercomparison of all data sets is given by the annual global, terrestrial and oceanic means, listed in Table 2. The highest total precipitation results from the forecast model and is about 8% larger than the highest value of the data sets based on conventional measurements, which have been corrected with regard to the systematic measuring error. Arpe /3/ studied the characteristics of the ECMWF forecast model results and found a tendency of the contemporary model to overestimate convective rainfall.

TABLE 2 Annual global, terrestrial and oceanic total monthly precipitation means
in [mm/year]   (L=landportion per grid area)

	global	continents (L > = 50%)	oceans (L < 50%)
Jaeger	985	840	1068
Legates(nc)	1034	804	1164
Legates(c)	1140	876	1284
Leemans/Cramer	-----	816	----
Hulme	-----	780	----
Schemm (1987),MSU	1323	804	1572
Schemm (1988),MSU	1277	840	1488
GPCC (1987)	1015	744	1152
GPCC (1988)	1025	780	1140
ECMWF (1987)	1205	912	1380
ECMWF (1988)	1230	924	1416

If the systematic error is not considered, the annual global means of the climatologies based on observational data are on the same level with relative differences of less than 3% of the total mean (1,015 mm). The overall mean of Legates' corrected data is about 10% larger than the result from his data as measured. A comparison of the two Legates data sets, separately for the climatic zonal belts and for the seasons, shows that the systematic gauge-measuring error is estimated to be highest and reaches its largest relative values with about 70% (absolute 15 mm/month) in the northern subpolar zone and more than 100% (absolute 30 mm/month) over Antarctica, in each case during the hemispheric winter season. In the tropics, the absolute error can also reach more than 10 mm/month, however, but the relative error is generally less than 5%.

The similarity of the climatological data sets and the model defect mentioned can also be seen by the comparison of the continental zonal means, plotted in Figure 1. For the northern summer as well as for the year, the ECMWF model yields the broadest ITCZ with highest precipitation, while the Hulme data indicate the lowest precipitation amounts in the tropics. Furthermore, there is a small shift in the location of the ITCZ between the model and the observation-based estimates in the northern winter season. In arid zones and in the northern hemisphere subtropics all results are in good agreement. In general, the largest differences between the observation-based climatologies are found in the southern hemisphere extratropics, what might be due to the sparse coverage with measured data.

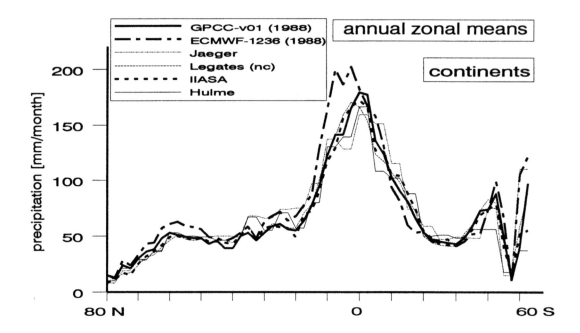

Fig.1. Latitudinal profile of zonal annual mean precipitation in mm/month for the continental areas, based on the climatological data sets of Hulme, Jaeger, Leemans/Cramer and Legates, and the results for the year 1988 for the GPCC and the ECMWF forecast results.

The WetNet PIP-1 (Barrett et al. /4/) global intercomparison of a large number of satellite-based estimates was carried out for monthly precipitation totals on a 2.5° grid and for a period of 4 months (August 1987 - November 1987). The Algorithm Intercomparison Program /2/ was set up for validation of new algorithms and was based on high resolution data from specific meso-scale regions and short periods within individual projects. Already finished are the studies AIP-1 over Japan with the data period June 1-30 1989 and July 15 - August 15 1989 and AIP-2 over U.K. and Western Europe for the period 1st February to 9th April 1991.

As a complementation, the GPCC is able to provide long-term large-scale comparisons. Up to now, the following satellite-based data are included: The operational infrared products of Janowiak /12/, the MSU-microwave-based estimates of Spencer /18/ and the SSM/I-microwave results of Chang (see Wilheit et al. /20/) and Ferraro /5/, /8/. The characteristics of the data sets are given in Table 3.

TABLE 3 Precipitation estimates based on satellite data

provided by	region	channels	algorithm	period
Chang	oceans; 5° lat/lon	microwave from SSM/I instrument 19 and 22 GHz	Wilheit et al./21/	Jul.1987 - Mar. 1994
Janowiak	tropics, land and oceans; 2.5° lat/lon	infrared from AVHRR instrument	Arkin /1/ /2/ Janowiak /12/ (GPI)	Jan. 1986 - Mar. 1994
Schemm	oceans; 5° lat/lon	microwave from MSU instrument 55 GHz	Spencer /18/	Jan. 1979 - Dec. 1992
Ferraro	oceans; 2.5° lat/lon	microwave from SSM/I instrument 19 and 22 GHz	Grody, Ferraro /8/ (emission)	July 1987 - Dec. 1991
Ferraro	oceanic and land areas; 2.5° lat/lon	microwave from SSM/I instrument 19,22 and 85 GHz	Ferraro et al. /5/ (scattering)	July 1987 - Dec. 1991

Janowiak represents the GSPDC which operationally provides precipitation estimates in the latitude belt 40°N to 40°S for land and ocean. The method is the GOES Precipitation Index /1/, where fractional areal coverage by cold clouds is used to estimate the tropical convective rainfall. Gaps in the geostationary satellites data (especially the lack of INSAT data) are filled by the observations of outgoing long-wave radiation (OLR) from NOAA polar orbiting satellites /12/. The main advantage of the information gained from IR data is that eight global images per day are available versus only two from microwave observations.

Chang with the PSPDC compiles the data of the Special Sensor Microwave Imager (SSM/I) from the Defense Meteorological Satellite Program. Precipitation is derived for the oceans between 50°N to 50°S using a radiation model and adjustment of histograms of calculated and observed brightness temperatures based on the 19 GHz and 22 GHz channels /20/.

Another way to detect precipitation over land from microwave radiation measurements can be using the scattering of radiation by larger ice particles present in the higher cloud layers, which depresses the 85 GHz signal. The scattering signal is also detectable over ocean, but mainly in rain systems that form over land and move offshore, or in intense tropical convective systems.

Over ocean, also frequency measurements at 19 and 22 GHz provide an emission signal caused by cloud and rain drops which increase the brightness temperature relative to the low emission of the sea-surface. Both the 19 GHz emission and the 85 GHz scattering technique, empirically tuned with coincident radar data, have been utilized in the SSM/I algorithms developed by Grody and Ferraro /5/, /8/.

The MSU monthly oceanic precipitation is estimated from Microwave Sounding Unit channels 1 (50.3 GHz), 2 (53.74 GHz) and 3 (54.96 GHz) data gathered by the TIROS-N satellites. Precipitation is diagnosed by Spencer's method: Cloud and rain water induced radiometric warming is indicated if the channel 1 brightness temperature exceeds a cumulative frequency distribution threshold of 15% after correction for airmass temperature determined from the channel 2 and 3 measurements.

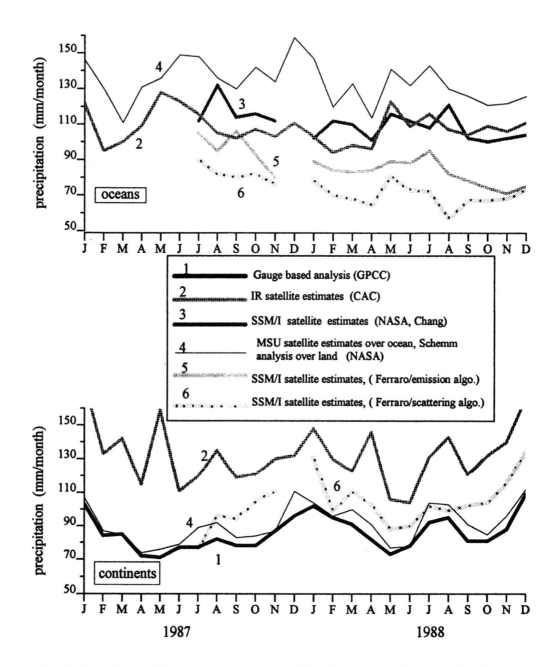

Fig. 2. Monthly precipitation averaged over a 60° wide latitude belt centered at the seasonal position of the ITCZ.

The annual course of the different data sets is here displayed for the tropics (Figure 2). For the oceans, the MSU-data indicate about 30% more precipitation, as it is estimated from the IR and the SSM/I data, both algorithms of Ferraro have lower results, while the NOAA/emission /8/ based approach seems to be more realistic. Over land, the NOAA/scattering /5/ based results of Ferraro is nearer to the analyses from ground-data than the IR-results.

The spatial characteristics are shown by means of zonal averaged monthly precipitation as shown in Figure 3 for autumn 1987. The largest differences between all data sets are found in mid-latitudes. The both raingauge-based terrestrial estimates, Schemm and the GPCC, are quite similar, even if the station density is very different.

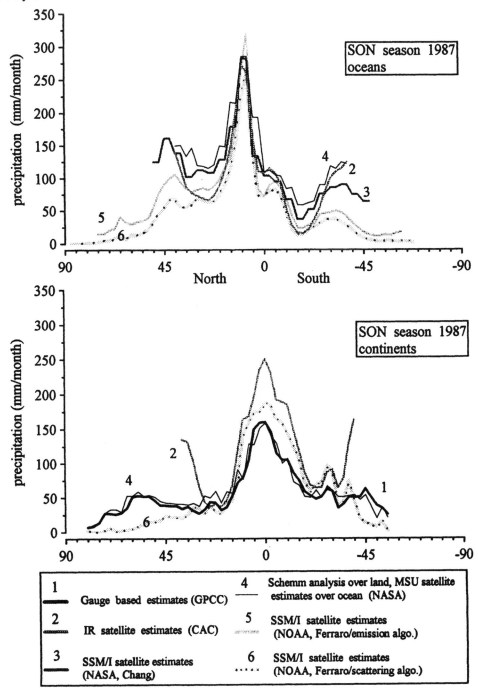

Fig. 3. Zonal averaged monthly precipitation for autumn 1987.

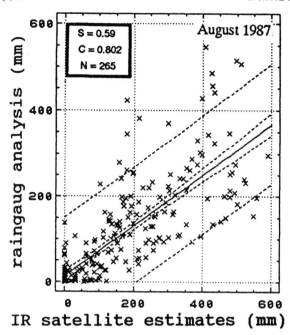

Fig. 4. Scatter diagram for grid-area related monthly precipitation for August 1987, analysis from rain-gauge measurements versus estimates from IR satellite images, r = correlation coefficient, s = linear regression slope, n = number of grid pairs (only less complicated grid cells are used, see text).

The estimates from satellite infrared images over land are gridwise compared to the results based on raingauge data. In order to avoid already known problematic cases, those grid-cells are not included, which contain high mountains (mean grid terrain level more than 1000 m msl), coasts (water-covered grid area of 50% or more) or less than two raingauges. In spite of these conditions, a widely spread scattering as well as a strong bias is calculated (Figure 4). Generally, IR-results overestimate the precipitation over the continents, and underestimate it over the oceans /6/. However, as Figure 5 shows, relationships diverging from this rule are found over the eastern coasts of the continents, where model results are higher, and over region of cold upwelling water, west of the continental shelves, where IR estimates are higher.

The spatial structure of the deficiencies of satellite based estimates has still to be analysed for other algorithms. Authors are invited to participate and to provide gridded results (preferably on the 2.5° grid and at first for the period 1986 to 1990 or part of it).

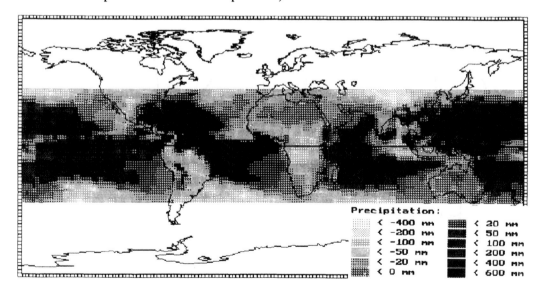

Fig. 5. Differences of annual precipitation derived from numerical weather forcast (ECMWF) and from infrared satellite observations (GPI-method) for 1987; [ECMWF-IR].

CONCLUSIONS AND OUTLOOK

The bias and the scattering between the compared precipitation climatologies based on conventional ground data are smaller than those between the compared satellite-based precipitation estimates. The differences of the ECMWF model results versus the raingauge-based climatological analyses are smaller than those of all the tested satellite based estimates versus reference data, as it is also shown by Barrett et al. /4/. In the satellite-data based analysis a large scattering of the precipitation estimates is seen, especially over oceanic regions. For precipitation estimates based on satellite data up to now no single alogorithm seems to be able to produce overall and everywhere reasonable precipitation amounts. A combination of multi-channel microwave and infrared information for precipitation estimates might lead to best-possible results.

But also the terrestrial raingauge-based analyses have to be improved. Since precipitation is highly variable both in space and time a station density of about 2-8 stations per 10,000 km^2, depending on the orographic and climatic conditions in the grid, is required to obtain area-average monthly precipitation on the 2.5° grid from point measurements with a mean relative error (due to spatial sampling) of less than 10% /17/. This amounts to a number of about 40,000 stations over the continents. Within its international scientific functions, the GPCC has acquired additional monthly precipitation data of more than 30,000 stations from about 100 countries. Besides, the raingauge-data have to be corrected with regard to the systematic measuring errors. A re-analysis based on all collected data and using improved methods is planned.

In order to derive digital global precipitation maps of the highest possible quality, the data of all suitable observation techniques have to be merged. An advanced method based on a error-dependant weighting-scheme and optimum estimation is in development /9/. As a precondition for applying such a method, the error ranges have to be derived separately for each observational data set considering the dependency on space and time. Investigations on the errors are in progress by the ongoing Algorithm Intercomparison Program and the WetNet Precipitation Intercomparison Project and GPCC error analysis studies.

Acknowledgement: The authors like to express their appricetion for the supply of gridded precipitation data, both climatologies and satellite based estimates. Thanks are primary addressed to P.Arkin, K.Arpe, A.Chang, R.Ferraro, M.Hulme, J.Janowiak, R.Leemans, D.Legates, and J.Schemm.

REFERENCES

1. P. A. Arkin, The relationship between the fractional coverage of high cloud and rainfall accumulations during GATE over the B-scale array. Mon. Weather Rev., 107, 1382-1387, (1979).

2. P. A. Arkin, J. E. Janoviak, and T. H. Lee, Atlas of Products from the Algorithm Intercomparison Project 1: Japan and Surrounding Oceanic Region June-August 1989. Published by the University Corporation for Atmospheric Research, Washington D.C., (1991).

3. K. Arpe, The hydrological cycle in the ECMWF short range forecasts. Dynamics of Atmospheres and Oceans 16, 33-60, (1991).

4. E. C. Barrett, D.R. Kniveton, C. Kidd, B. Motta, F. LaFontaine, M. Smith, and M. Goodman, Precipitation Intercomparison Project-1; PIP-1 results booklet: printed at Remoted Sensing Unit, University of Bristol,UK and at Marshall Space Flight Center, Huntsville, AL, USA, (1993).

5. R.R. Ferraro, N.C. Grody, and G.F. Marks, Effects of Surface conditions on rain identification using the SSM/I, Accepted for publication in Remote Sensing Reviews, (1994).

6. GPCC, Monthly precipitation estimates based on gauge measurements on the continents for the year 1987 (preliminary results) and future requirements. Rep.-No. DWD/K7/WZN-1992/08-1, Offenbach/Main, August 1992, 24 pp, (1992); (available from GPCC).

7. GPCC, Global area-mean monthly precipitation totals on a 2.5° grid for the year 1988 (preliminary results, derived from raingauge measurements, satellite observations and numerical weather prediction results). Rep.-No. DWD/WZN-1993/07-1, Offenbach/Main, July 1993, 20 pp, (1993); (available from GPCC).

8. N.C. Grody and R.R. Ferraro, A comparison of passive microwave rainfall retrieval methods - regional and global assessment. Proceedings of the 6th Conference on Satellite Meteorology and Oceanography, January 5-10, 1992, Atlanta, GA, American Meteorological Society, Boston, MA, 60-65, (1992).

9. G. J. Huffman, R. F. Adler, B. Rudolf, U. Schneider, and P. R. Keehn, Global precipitation estimates based on a technique for combining satellite-based estimates, raingage analyses and NWP model precipitation information. Submitted to J. Climate, (1994).

10. M. Hulme, An Intercomparison of Model and Observed Global Precipitation Climatologies. Geophys. Res. Lett.18, 1715-1718, (1991).

11. L. Jaeger, Monatskarten des Niederschlags für die ganze Erde. Bericht des Deutschen Wetterdienstes, Nr. 139 (Bd. 18), 33 pp and plates, (1976).

12. J. E. Janowiak, and P. A. Arkin, Rainfall variations in the tropics during 1986-1989, as estimated from observations of cloud-top temperature. J. Geophys. Res., 96, 3359-3373, (1991).

13. R. Leemans and W.P.Cramer, The IIASA Database for Mean Monthly Values of Temperature, Precipitation and Cloudiness on a Global Terrestrial Grid. Working Paper WP-90-41, Biosphere Dynamics Project, Internat. Inst. for Applied Systems Analysis, Laxenburg, Austria, 62 pp, (1990).

14. D. R. Legates, A Climatology of Global Precipitation. Publ. in Climatology 40 (1), Newark, Delaware, 85 pp, (1987).

15. B. Rudolf, Management and analysis of precipitation data on a routine basis. Proc. Internat. WMO/IAHS/ ETH Symp. on Precipitation and Evaporation (Eds. M. Lapin, B. Sevruk), 1, 69-76, Slovak Hydromet. Inst., Bratislava, Sept. 1993.

16. J. Schemm, S. Schubert and J. Terry, Estimates of monthly mean soil moisture for 1979-1989. NASA Technical Memorandum 104571, GSFC, Greenbelt, Maryland, (1992).

17. U. Schneider, B. Rudolf, and W. Rueth, The spatial sampling error of areal mean monthly precipitation totals analyzed from gauge measurements. Proc. Fourth Internat. Conf. on Precipitation "Hydrological and Meteorological Aspects of Rainfall Measurement and Predictability", Iowa City, Iowa, 80-82, (1993).

18. R. W. Spencer, Global oceanic precipitation dataset from the MSU 1979-1992. Journal of Climate, (1992).

19. WCRP, Global Precipitation Climatology Project - Implementation and Data Management Plan. WMO/TD-No.367, Geneva, June 1990, 47 pp. and appendices, (1990).

20. T. T. Wilheit, A. T. C. Chang, and L. S. Chiu, Retrieval of monthly rainfall indices from microwave radiometric measurements using probability distribution functions. J. Atmos. Ocean. Tech., 8, 118-136, (1990).

21. C. J. Willmott, C.M. Rowe and W.D. Philpot, Small-scale climate maps: A sensitivity analysis of some common assumptions associated with grid-point interpolation and contouring. The American Cartographer 12 (1), 5-16, (1985).

Pergamon

Adv. Space Res. Vol. 18, No. 7, pp. (7)63–(7)66, 1996
Copyright © 1995 COSPAR
Printed in Great Britain. All rights reserved
0273–1177/96 $9.50 + 0.00

0273–1177(95)00291–X

EFFECT OF SOIL MOISTURE AND CROP COVER IN REMOTE SENSING

D. Singh, P. K. Mukherjee, S. K. Sharma and K. P. Singh

Remote Sensing Laboratory, Department of Electronics Engineering,
Institute of Technology, Banaras Hindu University, Varanasi 221 005, India

ABSTRACT

An investigation with moist soil covered by narrow leaf crop at different stages of growth for establishing relation with scattering/reflection coefficient for both like polarizations (i.e. HH-pol and VV-pol) has been done. There is a linear relationship between crop covered soil moisture with scattering/reflection coefficient. Using linear relationship a theoretical model was proposed. The results from the model have been compared with the observed values. In this model, scattering/reflection coefficient was taken as dependent variable and percentage soil moisture (PSM) as independent variable. Regression parameters were obtained using the regression analysis which show that the brightness temperature at X-band depends significantly on the polarization that is chosen for receiving channel. The VV-pol is more sensitive to changes in the soil moisture, whereas, the contribution of crop morphology was more dominant near nadir for HH-pol. The calculated values show that the scattering increases as the PSM increases, whereas, observed values show no significant trend worth any conclusion.

INTRODUCTION

The successful application of radar remote sensing techniques to agricultural land use mapping is the understanding of the dependence of the back and forward-scattering coefficient (σ^o) from a vegetated scene.

The remote sensing application is also important for monitoring agricultural crops. Spaceborne and airborne platforms are helpful in providing repetative and consistent coverage of large areas. Among the specific applications of remote sensing in agriculture, currently under investigations, are crop discrimination and mapping, stage of growth determination and health of plants and their stress detection. In contrast, radar is independent of both the weather and the time of the day. The spaceborne and airborne radar, both imaging and non-imaging are in the process of recent developments even today, undergoing feasibility studies of the remote sensing techniques. Simultaneously, with the technical development of radar systems, experimental programme have been conducted to relate the radar back-scatter or forward-scatter to the phenology of agricultural crops of interest /1/.

Still the understanding of scattering mechanism is very limited. A rigorous theoretical solution to the electromagnetic scattering problem is not only very much hampered by the difficulties involved in defining plant geometry mathematically but would, if accomplished, probably be useful for prediction rather

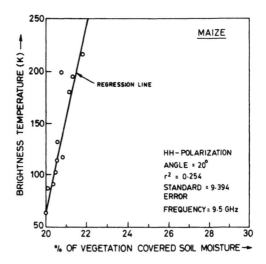

Fig. 1. Variation of brightness temperature with percentage of crop (narrow leaf) covered soil moisture (gravimetric) for HH-pol.

than sensing applications if it depended heavily on the widely variable plant geometry. Some very useful empirical model have been developed using regression analysis of the radar cross-section data on plant moisture, plant height and the moisture content of the underlying soil /2/. This type of analysis can sometimes be made more effective if it is based on functional relationship that reflect a hypothesis of the physical process involved. In this paper we have done a statistical regression analysis for narrow leaf crop i.e. maize covered soil.

THEORETICAL APPROACH

The value of scattering coefficient σ^0 was calculated as Sharma /3/;

$$\sigma^0 \text{ (dB)} = 10 \log_{10} \frac{[2|\rho_o|^2 \cot \phi_{az}/2 \csc \phi_{el}/2]}{[\sec (\theta-\phi_{el}/2) + \sec (\theta+\phi_{el}/2)]} \quad (1)$$

where $|\rho_o|^2$ = reflectivity of the target, θ, ϕ_{el}, and ϕ_{az} are the look or incidence, elevation and azimuth angles of the radar beam illumination. The details regarding the calculation of σ^0 is given by Sharma /3/.

Thus, knowing the values of look, elevation and azimuth angles of the radar beam illumination and the reflectivity of the target surface, we can calculate the scattering coefficient.

Theoretical model used here is;

$$\sigma^0 = m_g \frac{d (\sigma^0)}{dm_g} + I \quad (2)$$

where m_g is gravimetric soil moisture contents. $d (\sigma^0)/dm_g$ is slope and I is intercept.

RESULTS AND DISCUSSION

For the statistical regression analysis of narrow leaf crop (i.e. maize), the data have been taken from /3/.

As per the defination of botany, maize is classified as a narrow leaf crop. The crop attained a maximum average height of 96 cms during the experiments. As the crop leaves were narrow in shape but long in size, due to normal crop density which was kept rarer for better yield, the whole ground surface was not covered by this crop on which it was grown. So it must have had some effect from the uncovered background surface also.

Table 1 shows the regression analysis of the narrow leaf (i.e. maize) crop covered soil moisture for brightness temperature for HH-and VV-pols respectively. From Table 1, we observe that as the soil moisture increases, the brightness temperature increases for HH-pol, which is indicated by the positive sign of the slope. The slope also shows the sensitivity of the brightness temperature with the soil moisture, we note that the slope is not constant and varies for each incidence angle and therefore, does not give a good pattern. The sign of the slope is

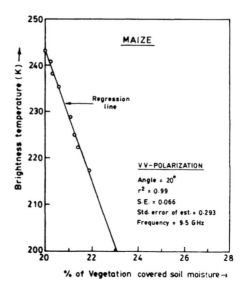

Fig. 2. Variation of brightness temperature with percentage of crop (narrow leaf) covered soil moisture (gravimetric) for VV-pol.

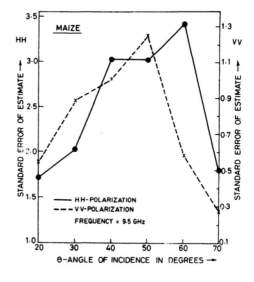

Fig. 3. Angular variation of standard error of estimate for scattering coefficient of crop covered soil.

negative for VV-pol in Table 1, which means that as the soil moisture increases the brightness temperature decreases. The slope of VV-pol is less than the slope of HH-pol. The value of coefficient of correlation is maximum 0.50 at the incidence angle of 20°, then its value decreases upto 50°. Further, its value increases again for HH-pol. The value of coefficient of correlation for VV-pol is 0.99 at incidence angle of 40°. The value increases upto 40° only and thereon it decreases. We find that the value of coefficient of correlation is very less for HH-pol in comparison to VV-pol. The variation of percentage of crop covered soil moisture with brightness temperature for HH-and VV-pols respectively are shown in Figures 1 and 2. We find that the observed value is greater than the calculated ones in both Figures. The observed value is scattered in the graph. It means that the crop covered soil does not give a good pattern. The solid line curves show the regression line in both the Figures for theoretically calculated values.

Table 2 shows the regression analysis for maize covered soil moisture for different angle with scattering coefficient (σ^O) for HH-and VV-pols respectively. As the soil moisture increases, the scattering coefficient decreases because the slope of regression line decreases for HH-pol (Table 2). However, as the percentage of soil moisture increases, scattering coefficient also increases, because slope is positive for VV-pol (Table 2). It means that both polarizations give different results. The slope which represent sensitivity to crop covered soil moisture decreases as the incidence angle increases (Table 2) for HH-pol. The sensitivity is maximum at 20°. There is very small change in the slope with angle, whereas, in Table 2 the sensitivity is negligible near nadir and its value reaches maximum at the higher angles. It means HH-pol has higher sensitivity at 20° and VV-pol has higher sensitivity at 70°. We also observe that the coefficient of correlation (r) varies from -0.18 to -0.59 for HH-pol, whereas, variation for VV-pol is from 0.05 to 0.099. It means that the value of 'r' is very less for HH-pol in comparison to VV-pol. We therefore infer that VV-pol gives better result in comparison to HH-pol. The maximum value of 'r' for HH-pol is 0.59 at incidence angle 20° and 0.99 at 70° VV-pol.

The angular variation of the standard error of estimate is shown in Figure 3. The value of standard error of estimate increases upto 60° and then decreases for HH-pol. The minimum value of standard error is 1.7 at 20°, whereas, this value increases upto 50° and then decreases for VV-pol. The minimum value is 0.29 at an incidence angle of 70°. We therefore conclude that VV-pol gives better result in comparison to HH-pol because of the fact that the value of standard error in VV-pol is very small as compared to that of HH-pol. However, results thus drawn have been compared with /4-7/ and found qualitatively similar in nature of variation. Due to difference in our system parameters and sensitivity as compared with others and the variation of climatic as well as environmental conditions, from place to place, we did not attempt to compare the magnitude.

For the sake of space conservation, the detailed results like; angular variation of coefficient of determination for brightness temperature for crop covered soil from narrow leaf (i.e. maize), incidence angles versus standard error curves, angular variation of stndard error of estimate for brightness temperature, variation between percentage of crop covered soil moisture (gravimetric) with scattering coefficient, angular variation of coefficient of determination (r^2) for crop (i.e.maize) covered soil moisture and the angular variation of the standard error for scattering coefficient could not be presented here, though they have been studied in detail.

CONCLUSIONS
1.   For both the polarizations, no significant trend for emissivity with crop covered soil moisture was inferable and hence no regression line can be drawn.
2.   Emissivity is more sensitive at incidence angles near nadir for HH-pol, whereas, the sensitivity is 40° for VV-pol.
3.   The brightness temperature in the X-band depends significantly on the polarization that is chosen for the receiving channel. VV-pol is found to be more sensitive to soil moisture than HH-pol.
4.   The contribution of crop was more dominant near nadir for HH-pol.

TABLE 1

Linear regression results of T_B Vs crop covered soil moisture (gravimetric) at different incidence angles for X-band.

Angle (Deg.)	Slope	Intercept	r^2	r	Std.Err.	Std.Error of Estm.	Pol.
20	5.48	63.29	0.25	0.50	9.39	26.93	HH
30	2.89	120.39	0.08	0.29	9.13	46.49	HH
40	3.65	89.39	0.04	0.20	88.87	27.78	HH
50	6.09	18.91	0.08	0.28	20.54	100.66	HH
60	2.97	112.08	0.12	0.34	17.96	39.81	HH
70	3.35	270.89	0.23	0.48	15.98	29.91	HH
20	-0.01	243.46	0.003	-0.05	1.94	8.63	VV
30	-3.34	327.65	0.99	-0.99	0.17	0.76	VV
40	-2.77	333.09	0.99	-0.99	0.06	0.29	VV
50	-0.90	301.92	0.37	-0.61	1.15	5.15	VV
60	-1.54	316.22	0.93	-0.96	0.42	1.89	VV
70	-1.22	308.86	0.86	-0.92	0.49	2.19	VV

TABLE 2

Linear regression results of σ^0 Vs crop covered soil moisture (gravimetric) at different incidence angles for X-band

Angle (Deg.)	Slope	Intercept	r^2	r	Std.Err.	Std.Error of Estm.	Pol.
20	-0.25	13.45	0.34	-0.59	0.34	1.71	HH
30	-0.14	10.77	0.09	-0.31	0.43	2.16	HH
40	-0.17	11.20	0.07	-0.26	0.63	3.17	HH
50	-0.17	10.52	0.07	-0.26	0.63	3.17	HH
60	-0.18	10.10	0.06	-0.25	0.72	3.57	HH
70	-0.05	0.22	0.03	-0.18	0.36	1.82	HH
20	0.006	5.28	0.003	0.05	0.12	0.54	VV
30	0.25	-1.51	0.93	0.97	0.06	0.89	VV
40	0.35	-6.09	0.70	0.84	0.23	1.02	VV
50	0.15	-3.44	0.12	0.34	0.42	1.86	VV
60	0.20	-7.51	0.70	0.84	0.13	0.58	VV
70	1.73	-38.65	0.99	0.99	0.002	0.29	VV

REFERENCES

1. F.T. Ulaby, T.E. Van Deventer, J.R. East, T.F. Haddock and M.E. Coluzzi, Millimeter-Wave Bistatic Scattering from Ground and Vegetation Targets, *IEEE Trans. Geosci. Rem. Sens.* 26, 229-243, (1988).

2. T.F. Bush and F.T. Ulaby, Radar Return from a Continuous Vegetation Canopy, *IEEE Trans. Ant. Prop.* AP-24, # 3, 269-276, (1976).

3. S.K. Sharma, Some Remote Sensing Applications of Interaction of Electromagnetic Waves with Materials, Ph.D. Thesis, Deptt. of Electronics Engg., Institute of Technology, Banaras Hindu University, Varanasi-221 005, India, 1991.

4. K.P. Singh and S.K. Sharma, Evaluation of Soil Moisture by Bistatic Microwave Remote Sensing, *Adv. Space Res.* 12, # 7, (7)69-(7)72, (1992).

5. T. Mo, T.J. Schmugge and T.J. Jackson, Calculations of Radar Backscattering Coefficient of Vegetation-Covered Soils, *Rem. Sens. Environ.* 15, 119-133, (1984).

6. T.J. Jackson, T.J. Schmugge and P.O'Neill, Passive Microwave Remote Sensing of Soil Moisture from an Aircraft Platform, *Rem. Sens. Environ.* 14, 135-151, (1984).

7. T. Mo, T.J. Schmugge and J.R. Wang, Calculations of the Microwave Brightness Temperature of Rough Soil Surfaces: Bare Field, *IEEE Trans. Geosci. Rem. Sens.* GE-25, 47-54, (1987).

 Pergamon

Adv. Space Res. Vol. 18, No. 7, pp. (7)67–(7)71, 1996
1995 COSPAR
Printed in Great Britain
0273–1177/96 $9.50 + 0.00

0273–1177(95)00292–8

TOWARDS SATELLITE-DERIVED GLOBAL ESTIMATION OF MONTHLY EVAPOTRANSPIRATION OVER LAND SURFACES

G. Gutman* and L. Rukhovetz**

* *Satellite Research Laboratory, NOAA/NESDIS, Washington, DC 20233, U.S.A.*
** *Research and Data System Corporation, Greenbelt, MD 20779, U.S.A.*

ABSTRACT

The NOAA AVHRR provides daily global spectral measurements of reflected and emitted radiation from Earth. These data contain information on land characteristics, which could be utilized to derive evapotranspiration. A global land surface data base has been developed from 9-year time series of multispectral AVHRR data consisting of Channels 1,2,4, and 5. The data are sampled in time and space and mapped to $(0.15°)^2$ grid on a weekly basis. Processing includes proper calibration, cloud screening, monthly averaging, interpolation and smoothing.

It is shown, in the present study, that when evaporation exceeds precipitation the soil moisture availability parameter ß correlates well with an index of vegetation activity NDVI, derived from the visible and near-IR AVHRR measurements. If potential evaporation Ep is estimated, e.g. from Thornthwaite's empirical formula that uses only air temperature as input, then the actual evapotranspiration, E, can be calculated using the NDVI-derived ß as a fraction of Ep. In this study, the NDVI-ß relationship is determined using a previously developed climatology of ß based on hydrologic balance calculations. An example of NDVI-ß relationship for July over the globe is given.

INTRODUCTION

Mintz and Walker /1/ described a climatology of land surface evapotranspiration and soil moisture that had been produced more than a decade ago in /2/ and that has become a standard global data set for use in experiments with General Circulation (GCM) and Numerical Weather Prediction (NWP) Models. The last section in /1/ suggests a way to bypass root zone soil moisture calculation and derive land surface energy and water budget components using the Normalized Difference Vegetation Index (NDVI), derived from measurements by the Advanced Very High Resolution Radiometer (AVHRR).

The main premise for using the AVHRR observations is that the ratio of actual to potential evapotranspiration, the so-called ß-parameter, or root zone soil moisture availability, is related to canopy resistance which, in turn, was shown to be directly related to the ratio of near-IR to visible reflectances /3/, i.e. a function of NDVI. The governing equations and the preliminary results were described in /4/ but never published in a reviewed journal after Professor Mintz's death. The calculated fields were on a 4°x5° latitude/longitude grid and used the 1982-1984 Global Vegetation Index (GVI) data, which are strongly cloud contaminated. The present note is an attempt to follow the idea to use satellite observed "greenness" to estimate transpiration which is presumably the dominant component of evapotranspiration. This approach is tested on a monthly basis with enhanced NOAA GVI products /5/.

BACKGROUND

Derivation of evapotranspiration, E, from satellite remotely sensed visible and IR data has been on the scientific agenda for over 15 years. The thermal IR methods assume the knowledge of the surface net radiation flux and certain meteorological parameters as input to E- parameterizations (e.g. /6/). These studies derive E as a residual from the surface energy balance. Other methods that use various E-parameterizations involve satellite derived NDVI in its relationship with roughness, Leaf Area Index, fractional vegetation amount, and canopy stomatal resistance (e.g. /7, 8, 9, 10, 11/). Recently, many of these studies have been based on the analysis of the relationship between NDVI and satellite observed temperatures. It is implied that such a relationship should be used to quantify soil moisture availability (or canopy resistance) rather than NDVI alone. Most of these regional studies, however, assume that the surface net radiation flux is known. Some of the models used involve calculation of the soil water balance and need meteorological information as a necessary input. This approach deserves further investigation on a global scale. In the present study, however, we focused on implementing Mintz and Walker's /4/ proposal, which considers NDVI alone as a driving factor that characterizes the root zone soil moisture availability. We, thus, are interested in investigating the information content of the derived NDVI, whereas future refinement to present formulation can potentially be made with an inclusion of a thermal factor.

It has been shown that potential evapotranspiration on a long-term basis can be well approximated by the net radiation flux at the earth's surface , which, however, is not yet readily available on a global scale except for special measurements at selected sites. Using observations from several stations in different climates and comparing results from different methods, Mintz and Walker /1/ justified their use of Thornthwaite's method for calculating potential evapotranspiration, E_p, which uses screen-level air temperature and length of daylight hours, h, as the only input, and modified that formulation by adjusting the "true" temperature in conditions of soil moisture stress to the one that would be observed under potential conditions.

The ß-parameter is the coefficient of transpiration usually calculated as a function of the ratio of the root-zone moisture to the root-zone storage capacity. An innovative approach was proposed in /4/ to make this parameter dependent on the observed greenness as manifested by NDVI. The present study is making a next step towards using ß derived from NDVI. The goal is to follow /4/ in estimating monthly actual evapotranspiration without involving soil moisture calculations, but using a better resolution of $(0.15°)^2$ grid of the NOAA GVI and employing the enhanced GVI monthly products, obtained after proper calibration, screening of cloud contamination, monthly averaging, interpolation and smoothing (/5/). These products have been developed from a global land surface 9-year (1985-1993) weekly time series of AVHRR multispectral measurements including visible, near-infrared and thermal infrared data and the corresponding viewing and sun geometry. Monthly maps of 5-year means of NDVI were calculated and are used in the present study.

PRELIMINARY RESULTS

Monthly evapotranspiration is estimated in this study as:

$$E=ßE_p \qquad (1)$$

where monthly potential evapotranspiration, E_p, is calculated following /1/, using Thorntwaite's formulation with monthly mean climatological screen-level air temperature, T_a, as input. The values of ß from /1/ is used to find a relationship with NDVI on a global basis for July data:

$$ß= a + b*NDVI \qquad (2)$$

The NDVI data were restricted to solar zenith angles < 60° and Brightness Temperature Difference between Channels 4 and 5 (BTD) ≤ 3K. This screening eliminates observations at low sun angle and high water vapor amount, both factors introducing errors in NDVI. Also, to remove high altitude areas only those data points with $T_a>10°C$ were retained. The remaining data were averaged for 1°x1° areas and merged with ß produced in /1/ from hydrologic balance calculations on a 4°x5° grid. The resulting scattegram is shown in Fig. 1. It was noticed that the ß-NDVI relationship breaks down for the data points with small or no moisture deficit manifested by P-E, where P is monthly precipitation. Those points having E-P<1cm but with a significant amount of rain (P>1cm) are shown as boxes and the rest, i.e. those points having E-P>1cm or with a negligible rain amounts (P<1cm), are shown as circles. The circles produce a ß-NDVI relationship similar to that reported in /12/, in which satellite derived NDVI were compared with measurements of E/E_p for the growing season in Australia.

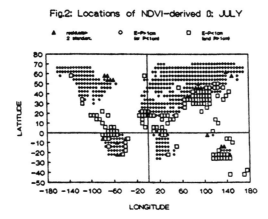

Fig.1: ß-NDVI relationship: JULY
Mintz's ß vs 1 deg mean NDVI

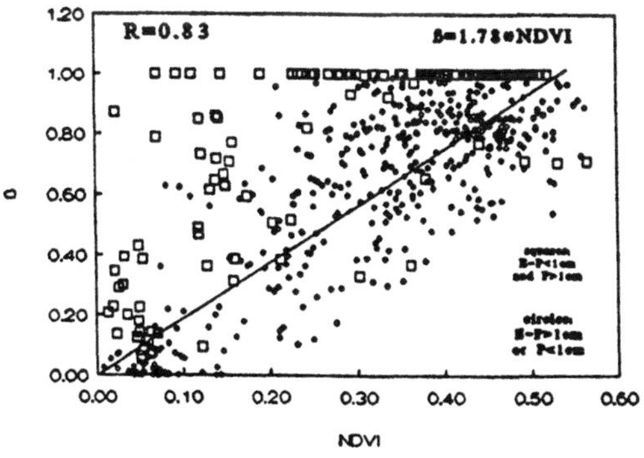

Fig.2: Locations of NDVI-derived ß: JULY

It can therefore be preliminarily concluded that the proposed method is limited to the situations when rain amount is smaller than evaporation, a typical case being the growing season in the middle latitudes. For July data, a reasonably strong linear relationship was found between ß and NDVI, with correlation coefficient 0.83. Addition of temperature as a predictor increases this coefficient to 0.87. The presently used value of 1cm for E-P to restrict the applicability of the method is yet to be tested for sensitivity. The maximum NDVI value of 0.58 is higher than 0.4 used in /4/ because of recalibration and cloud screening applied. The ß-NDVI relationship deserves further investigation and refinement, especially in view of the forthcoming enhancements of the NDVI data by atmospheric corrections that generally increase the NDVI for vegetated surfaces. Note that the presently obtained relationship is not generic since it is based on the July data alone.

Locations of the NDVI-derived ß are shown in Fig. 2. Data points with E-P<1cm and E-P>1cm are shown as boxes and circles, respectively. The residuals greater than 2 standard errors, indicated by black triangles, are concentrated in coastal areas and high elevation areas north of India.

The equations (1)-(2) were used to produce a global monthly climatology of E for July from the 5-year climatologies of T_a and NDVI (1985-1991). The year 1988 -- the last year of NOAA-9 -- was excluded from the base period because AVHRR observations were collected when the sun was low as a result of satellite orbit drift. The T_a fields were obtained from monthly data at stations over the globe available at NOAA Climate Analysis Center. The data were edited for quality to remove spurious observations and interpolated into $(0.15°)^2$ grid used for the NDVI fields. For the areas of applicability (E-P>1cm), discrepancies of the derived E values with those reported in /1/ are due to residual regression errors and small differences in the calculated E_p.

CONCLUSIONS

Preliminary results on global mapping of evapotranspiration suggest that the use of satellite derived vegetation indices makes it possible to bypass soil moisture calculations when deriving surface energy fluxes. If absolute values of soil moisture content are not needed, this method provides an easy way to map latent heat flux on a global scale. The current method is attractive because of its simplicity: a very simple model with only two input variables, T_a and NDVI. Preliminary results indicate that the present approach is unsuitable for the cases with monthly moisture P-E exceeding some threshold. More analysis using other months are necessary to confirm this conclusion. Preliminary studies of data for other months indicate that the ß-NDVI relationship is not universal, varying with month and climate region. Validation and comparisons will be made using off-line GCM and NWP model calculations. The next logical step will be to make an attempt to derive E from satellite data alone, i.e. without involving conventional observations. Together with satellite-derived precipitation, the derivation of E from AVHRR will bring us closer to the aim of GEWEX -- to better understand the hydrologic cycle and its variability on the global scale.

Acknowledgement

Thanks go to Dan Tarpley for encouragement and discussions on this study. Programming assistance at the preliminary stage of this work by Dan Sullivan of Research Data Systems Corporation is appreciated. The work has been supported by the NOAA Climate and Global Change Program.

REFERENCES

1. Y. Mintz and G.K. Walker, Global fields of soil moisture and surface evapotranspiration derived from observed precipitation and surface air temperature, *J. Appl. Met*, 32, 1305-1334. (1993)

2. Y. Mintz and Y. Serafini, Global fields of soil moisture and surface evapotranspiration, *Research Reviews-1980/81*, 178-180 (1981).

3. P.J. Sellers, Canopy reflectance,photosynthesis and transpiration. *Int. J. Rem. Sens.*, 6, 1335-1372 (1985).

4. Y. Mintz and G.K. Walker,Land surface energy and water budjets derived with NDVI. *Proc. Workshop on the 'Use of Satellite-derived Vegetation Indices in Weather and Climate Prediction Models*. Camp Springs, MD, February 26-27, pp.113 (1990)

5. G. Gutman, D. Tarpley, A. Ignatov, S. Olson, Global land products from AVHRR: The enhanced NOAA GVI dataset, *Bulletine Amer. Met. Society*, in press (1994).

6. B. Seguin, J.-P. Lagouarde and M. Savane, The assessment of regional crop water conditions from meteorological satellite thermal infrared data, *Remote Sens. Env.*, 35, 105-148 (1991).

7. J.D. Tarpley, Monthly evapotranspiration from satellite and convential meteorological observations. *J. Climate*, 7, 704-713 (1994).

8. J.C. Price, Using spatial contex in satellite data to inter regional scale evapotranspiration. *IEEE Trans. Geosc. Remote Sens.*, 28, 940-948 (1990).

9. J.C. Price, Estimating Leaf Area Index from satellite data. *IEEE Trans. Geosc. Remote Sens.*, 31, 727-734 (1993).

10. S.W. Running, Computer simulation of regional evapotranspiration by integrating landscape biophysical attributes with satellite data, 359-370 in *Land Surface Evaporation*, ed. T.J. Schmugge and J.-C. Andre, pp.424, (1991).

11. T.N. Carlson,Recent advances in modeling the infrared temperature of vegetation canopies. 349-338 in *Land Surface Evaporation*, ed. T.J. Schmugge and J.-C. Andre,(1991), p.424.

12. R.C.G. Smith and B.J. Choudhury, Relationship of multi-spectral satellite data to land surface evaporation from the Australian continent. *Int. J. Remote Sens.*, 11, 2069-2088.

Pergamon

Adv. Space Res. Vol. 18, No. 7, pp. (7)73–(7)76, 1996
Copyright © 1995 COSPAR
Printed in Great Britain. All rights reserved
0273–1177/96 $9.50 + 0.00

0273–1177(95)00293–6

LOW-LEVEL WIND FIELDS FROM GEOSTATIONARY SATELLITES AND WATER VAPOUR TRANSPORT

A. Ottenbacher, J. Schmetz and K. Holmlund

*European Space Agency (ESA), European Space Operations Centre (ESOC),
64293 Darmstadt, Germany*

ABSTRACT

Low-level cloud-motion winds are operationally derived from successive Meteosat infrared (IR: 10.5-12.5 μm) images. The operational cloud motion winds have been augmented by the tracking of cloud features in the visible (VIS: 0.4 - 1.1 μm) channel. The advantage of the VIS channel is a significantly better spatial resolution of 2.5 km × 2.5 km as compared to 5 km × 5 km in the IR.
The higher spatial resolution improves the tracking which results in a larger number of low level cloud motion winds. The observed low-level wind fields are combined with the mixing ratios at 850 hPa over the Atlantic ocean from short term forecasts in order to compute zonal and meridional indices of the water vapour transport over the ocean.

INTRODUCTION

Global observations of atmospheric wind fields are potentially the most important data in the analysis for numerical weather prediction (NWP) /1/. Direct wind observations are indispensable at low latitudes where winds cannot be inferred from the mass field. Wind observations from satellites also constitute the sole source of wind data over wide regions of the Southern Hemisphere.

At the European Space Operations Centre (ESOC) cloud-motion winds derived from infrared images (IR-CMWs) of the European geostationary Meteosat satellites were produced four times a day. While low-level cloud winds are routinely derived from the IR channel for use in numerical weather prediction little effort has been made so far to exploit the high-resolution visible images of the purpose of wind derivation. Recent progress in that area at ESOC has the potential to contribute to an improved understanding of low-level water vapour transport over the ocean. The lack of understanding concerning atmospheric water vapour provided the incentive to conduct GEWEX /2/. It is the purpose of this paper to illustrate how developments and improvements of operational satellite products can make an important contribution to climate research.

CLOUD-MOTION WINDS FROM METEOSAT SATELLITE IMAGES

The geostationary Meteosat satellites observe the earth with an imaging radiometer in three channels: in the solar spectrum (VIS) between 0.4 and 1.1 μm, in the infrared window region (IR) between 10.5 and 12.5 μm, and in the water vapour (WV) absorption band between 5.7 and 7.1 μm. Images are taken at half hourly intervals and the spatial sampling at the subsatellite point corresponds to 2.5 km × 2.5 km for the VIS, and 5 km × 5 km in the IR and WV channels.
The operational derivation of cloud-motion winds from successive IR images began in the second half of the 1970's and since then has been continuously refined and improved /3/. The tracking of low level cloud features in consecutive VIS images has just recently been implemented.

Figure 1 shows an example of the VIS-CMW product with a total of 2855 low level wind vectors. It was obtained from one production run for 1100 UTC 3 June 1994, after the application of a local consistency check, which removes wind vectors that disagree in speed, direction or height with the surrounding wind vectors. The derivation of low level VIS-CMWs has been restricted to sea areas only. This is because the tracking of the low level VIS-CMWs uses mainly stratus and stratocumulus, which usually cover large areas over sea and not over land. The height assignment of the low level VIS-CMWs is based on the IR channel, which assigns a equivalent blackbody temperature (EBBT) to the identified cloud top.

The quality of the VIS-CMWs is monitored routinely by comparisons with the wind forecast of the European Centre for Medium-Range Weather Forecast (ECMWF) in Reading, UK. First results indicate that the quality of the VIS-CMWs that are only checked for local consistency is worse than the quality of the operational low-level IR-CMWs, which have already been quality checked against the ECMWF wind forecast. The root mean square (RMS) error of the vector difference between the CMWs and the wind forecast, for example, is about 4.75 m/s for the VIS-CMWs and about 3.5 m/s for the IR-CMWs. However, the higher vector difference may also be due to the fact that the satellite VIS winds resolve features at a higher spatial resolution.

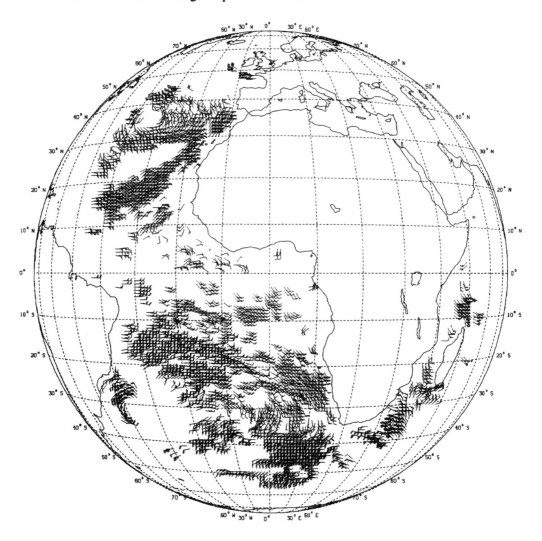

Fig. 1. Low-level cloud-motion winds from Meteosat VIS images for 1100 UTC 3 June 1994.

DERIVATION OF A WATER VAPOUR TRANSPORT INDEX

In order to illustrate the usefulness of the low-level VIS winds for climate and NWP applications an index of water vapour transport is studied in the following manner :

For a period of three days, from 3rd to 5th of June 1994, the low level VIS-CMWs and the ECMWF wind, temperature and humidity forecast was taken to derive two different indices for the water vapour transport in the lower atmosphere. Firstly, the three ECMWF wind forecasts for 850 hPa and 1200 UTC were merged together and the averaged wind vectors were multiplied with the similar averaged ECMWF mixing ratio forecast for the same time period. Secondly, the low-level wind vectors of the three VIS-CMW production runs for 1100 UTC were merged together and then also multiplied with the corresponding 3-day average of the ECMWF mixing ratio forecast. The two resulting indices for the water vapour flux in the lower atmosphere were then divided into a meridional and a zonal component. Figure 2 gives these two components for the water vapour transport indices derived from ECMWF data only (ECMWF/ECMWF) in the upper half of the chart and the corresponding components derived from VIS-CMWS and ECMWF mixing ratio (VIS-CMW/ECMWF) in the lower part of the chart. Figure 2 suggests that the meridional component of the water vapour transport indices shows larger differences than the comparison of the zonal components.

A comparison of the meridional components of the water vapour transport indices shows that the use of satellite winds leads to a more pronounced water vapour flow in the trade winds feeding the tropical deep convection. Significant differences are also observed in the South Atlantic off the coast of South America. While this latter discrepancy is probably due to inconsistencies in the satellite winds because of contamination by high-level cloud winds, the differences in the trades might indicate small deficiencies in the ECMWF wind field. The absolute maxima of the water vapour transport indices off the coast of Tansania and Kenia and over Somalia are difficult to compare, since no VIS-CMWs are derived over land. Therefore only the marine part of this index maximum can be determined by the VIS-CMW/ECMWF water vapour transport index.

Comparing the zonal components of the two water vapour transport indices shows similar maxima over the same geographical regions. The only striking difference between the two water vapour transport indices is the shape of the local index maxima over the northern part of the South Atlantic, where the ECMWF/ECMWF water vapour transport index resolves the two maxima more clearly than the VIS-CMW/ECMWF index. The missing zonal maximum of the VIS-CMW/ECMWF water vapour transport index over Somalia is again due to the fact, that no VIS-CMWs have been derived over land.

CONCLUSIONS

A method has been developed to derive low-level cloud-motion winds from high-resolution visible images from Meteosat. The better spatial resolution of the VIS channel leads to an improved resolution of mesoscale wind features. The quality impact on NWP of this new data source will be tested this summer in a joint campaign between ESOC and ECMWF.

Zonal and meridional water vapour indices have been computed from the satellite winds and forecast mixing ratios on the one hand and from forecast data on the other hand. Differences are seen between the two data sets. Although parts of the differences might be due to shortcomings in the satellite data it is suggested that the new VIS wind data have a good potential to improve weather forecasts and our capability to correctly model water vapour transports.

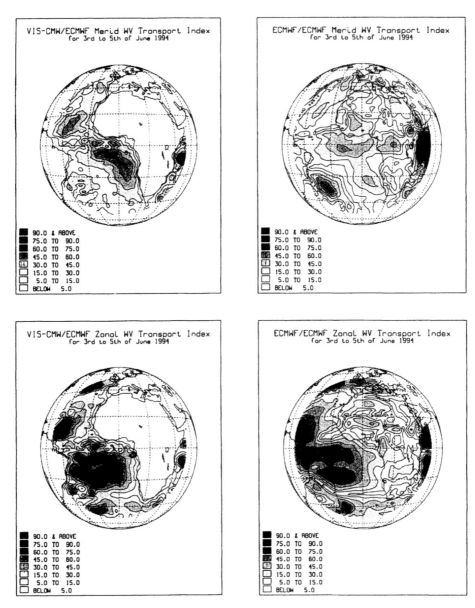

Fig. 2. Zonal and meridional component of the water vapour transport indices for the lower atmosphere from 3rd to 5th of June 1994. Index values are given in m s^{-1} g kg^{-1}.

REFERENCES

1. Baker, W.E., 1991 : Utilization of satellite winds for climate and global change studies. *Proc. NOAA Conf. on Operational Satellites: Sentinels for the Monitoring of Climate and Global Change, Global Planetary Change*, **4**, 157-163, [special issue]

2. Global Energy and Water Cycle Experiment, 1992, *Report of the First GEWEX Temperature/Humidity Retrieval Workshop, Greenbelt, U.S.A., 23-26 October 1992*, WMO/TD-NO. 460.

3. Schmetz, J., K. Holmlund, J. Hoffman, B. Strauss, B. Mason, V. Gaertner, A. Koch and L. van de Berg, 1993: Operational Cloud-Motion Winds from Meteosat Infrared Images. *J. Appl. Meteor.*, **32**, 1206-1225.

Chapter 2
Ocean and Land Productivity as Derived from Satellite Data
(Mtg A4)

PREFACE

The emphasis of this session organized by Dr. Richard Dugdale was on the use of satellite data to provide estimates of land and ocean productivity using models to obtain quantitative results. The session was divided into four sections: 1) land and ocean nutrients chaired by Dr. J-P. Malingreau, 2) ocean nutrients and upwelling chaired by Dr. T. Platt, 3) primary production and ocean color productivity models chaired by Dr. R. Dugdale and 4) ocean models chaired by Dr. A. Gabric. The land productivity systems were reviewed in the first section, along with a paper using aerial photography to assess land salinization. In section 2, papers were presented showing that it was possible to estimate surface nutrient concentrations in the equatorial Pacific upwelling area from sea-surface temperatures obtained from the satellite-borne Advanced Very High Resolution Radiometer (AVHRR) sensor. In the same section a model for productivity in a coastal upwelling region using AVHRR surface temperatures was presented. The need to use a regional approach in these studies was a common theme. The use of AVHRR temperatures was an alternative approach to the more common use of ocean color data from satellites to estimate primary production in the sea. This latter approach was the theme for section 3 where both approaches (using AVHRR and ocean color) were reviewed for the state-of-the-art and application to specific coastal systems presented. In section 4, bio-optical models and their linkage to satellite-derived ocean color data and ocean productivity estimates were presented. A data base system for examination of world ocean pigment distributions was also presented and the use of remote sensing to predict dimethylsulfide flux from ocean to atmosphere was considered.

Pergamon

Adv. Space Res. Vol. 18, No. 7, pp. (7)81–(7)89, 1996
Copyright © 1995 COSPAR
Printed in Great Britain. All rights reserved
0273-1177/96 $9.50 + 0.00

0273-1177(95)00951-5

SILICATE VERSUS NITRATE LIMITATION IN THE EQUATORIAL PACIFIC ESTIMATED FROM SATELLITE-DERIVED SEA-SURFACE TEMPERATURES

F. P. Wilkerson and R. C. Dugdale

Department of Biological Sciences, University of Southern California, University Park, Los Angeles, CA 90089-0371, U.S.A.

ABSTRACT

Productivity in the equatorial Pacific may be limited by a variety of factors including the availability of either nitrate (NO_3) or silicate ($Si(OH)_4$). The large size of this oceanic region makes it difficult to obtain a synoptic view of the nutrient field in traditional ways to assess NO_3 or $Si(OH)_4$ limitation. However, since concentrations of both nutrients are correlated with temperature, we have been able to employ remotely-sensed sea-surface temperatures (SST) from AVHRR to estimate the variability of surface NO_3 and $Si(OH)_4$ concentrations, and the ratio of one nutrient to the other which can be used to evaluate nutrient limitation in a Liebig sense. Using zonal transects along the equator, nutrient concentrations obtained from AVHRR-SST data from 1987 and 1988 show $Si(OH)_4$ values to be typically less than NO_3 and $NO_3:Si(OH)_4 > 1$, indicating $Si(OH)_4$ to be the limiting nutrient and the equatorial Pacific to be a HNLSLC (High Nitrate, Low Silicate, Low Chlorophyll) area. However, in warmer climatic conditions, i.e. during 1987--an El Niño year and in the western part of the equatorial swath, NO_3 falls to undetectable levels and is most likely the more limiting macronutrient. Data from 1988 for 15°N to 15°S, 180°W to 90°W shows maximal concentrations of both nutrients to occur to the south of the equator with the ratio values indicating $Si(OH)_4$ limitation to the south and NO_3 limitation to the north of the equator.

INTRODUCTION

Large and significant regions of the oceans are characterized by values of chlorophyll concentration and primary productivity unexpectedly low in view of apparent nutrient sufficiency. These have been designated HNLC, i.e."High Nutrient, Low Chlorophyll" /1/ and include such areas as the eastern equatorial Pacific, the region offshore of the Peru and northwest Africa coastal upwelling areas, and the Southern Ocean. In making these designations, the nutrient considered was NO_3. However, a re-examination of the nutrient status of these HNLC regions revealed that another macronutrient, $Si(OH)_4$, occurred in concentrations very much lower than those of NO_3. Consequently those HNLC regions designated above were more accurately described as "High Nitrate, Low Silicate, Low Chlorophyll" areas /2/. The causes of the low $Si(OH)_4$ condition of these waters appears to be the result of the regeneration of $Si(OH)_4$ occurring deeper in the water column than for nitrogen, and the differential export of $Si(OH)_4$ through zooplankton grazing. Whatever the underlying processes bringing about low $Si(OH)_4$ relative to NO_3, these HNLSLC regions clearly suffer from actual or incipient $Si(OH)_4$ limitation of diatom productivity and since this element of the primary productivity system is crucial to vertical flux processes (e.g. /3/) attention must be given to silicate processes in addition to nitrogen flux, in these large areas of the ocean.

One such region is the equatorial Pacific, where time scales of change are rapid due to the presence of waves with about 20-day period /4, 5/ that result in considerable north-south variability. The vast size of this area precludes obtaining a synoptic picture of nutrients and productivity from shipboard or moorings. Satellite observations, however provide synoptic data, relatively extensive in time and space. We have previously used AVHRR-derived SST data to map surface NO_3 concentrations of the equatorial Pacific /6/. In this communication, we use AVHRR-derived SST to estimate $Si(OH)_4$ in addition to NO_3 concentrations, and calculate the ratio of the two nutrients to evaluate the nutrient status and variability of the equatorial Pacific.

METHODS

Shipboard data of temperature and nutrients for the equatorial Pacific collected between May to December 1988 during EPOCS (Equatorial Pacific Ocean Climate Studies) cruises as part of the TOGA (Tropical Ocean Global Atmosphere) Program, were obtained from Dr. F. Chavez at the Monterey Bay Aquarium Research Institute. These cruises (13 to 25 May, 2 June to 5 July, 17 to 25 July, and 23 October to 3 December) were made within the area 15°N to 15°S, 180°W to 90°W. The data were divided into latitudinal sectors and the linear (for NO_3) or exponential (for $Si(OH)_4$) regressions calculated for nutrient versus temperature (Table 1). These regressions compare favorably to those calculated by others for this area (e.g. /7, 8, 9/).

TABLE 1 Nutrient Versus Temperature (T) Regressions

Latitude	$[NO_3] = b + m * T$			$[Si(OH)_4] = b * exp(m * T)$			
	m	b	r^2	m	b	r^2	n
15°N to 5°N	-2.06	55.47	0.97	-0.17	213.2	0.98	451
5°N to 1°N	-1.97	53.86	0.98	-0.16	176.2	0.96	391
1°N to 1°S	-1.25	36.92	0.91	-0.15	141.6	0.95	649
1°S to 5°S	-1.80	51.78	0.95	-0.15	149.7	0.92	124
5°S to 15°S	-1.66	52.25	0.94	-0.13	130.2	0.94	14

Weekly averaged SST data from the University of Miami/Rosenstiel School of Marine and Atmospheric Sciences MCSST were obtained from the NASA Ocean Data System /10/ for the years 1987 and 1988. Data (with a spatial resolution of 18 km) were extracted for our study area-a box 15°N to 15°S, 180°W to 90°W. The weekly averaged AVHRR images that were used in this study were selected to illustrate the upwelling (September) and non-upwelling (April) seasons /11, 12/ during El Niño (1987) and La Niña (1988; /13, 14/) conditions (Table 2). The appropriate latitudinal sector regressions (Table 1) were applied to pixel values of SST with a linear interpolation between the values of intercept and slope for each line of an image /6/. These NO_3 and $Si(OH)_4$ data were ingested into SEAPAK software package /15/ and pseudocolor images produced for NO_3, $Si(OH)_4$ and $NO_3:Si(OH)_4$ ratio. Zonal sections from 180°W to 90°W were obtained by averaging the three north to south pixels at the equator from the images.

TABLE 2 Weekly Averaged AVHRR Images Used

Julian Date	Week Number	Month	Conditions
87119	16	April	non upwelling, El Niño
87273	36	September	upwelling, El Niño
88111	16	April	non upwelling, La Niña
88237	34	September	upwelling, La Niña

The nutrient in minimal supply will be the nutrient that limits productivity /16, 17/. The ratio of $NO_3:Si(OH)_4$ can be used as a stoichiometric indicator of potential $Si(OH)_4$ versus NO_3 limitation, with the assumption that the utilization ratio of $NO_3:Si(OH)_4$ is 1:1 for diatoms /18/. Values of $NO_3:Si(OH)_4$ greater than one suggest that $Si(OH)_4$ will be limiting and NO_3 in greater supply, whereas the converse is likely when the ratio is less than one (i.e. $NO_3:Si(OH)_4 = 0.5$), with NO_3 less available than $Si(OH)_4$.

RESULTS

The equatorial values of nitrate and silicate concentrations from 180°W to 90°W obtained from the four SST images show NO_3 concentrations to range from zero (i.e. below Autoanalyzer detection) to almost 14 µM and $Si(OH)_4$ to be rarely below 2 µM and reaching up to 9 µM (Figure 1). The higher concentrations tend to occur east of 120°W with maximal values usually at about 92°W and low nutrient values to the west. Interestingly in all the periods there appears to be a drop in nutrient concentrations at 100°W.

The nutrient concentration values extracted from the April images (Figures 1a, c)show little east-west variability with low values (NO_3 and $Si(OH)_4$ mostly 2 µM or less) along the entire equator during the El Niño year of 1987 (Figure 1a), consistent with the lack of equatorial upwelling at this time, when easterly winds along the equator are weak /12/. The same trend is seen in 1988 with $Si(OH)_4$ staying almost constant at 3-4 µM with a few values slightly higher at 106°W and 92°W, where NO_3 concentrations also reach a maximum of 7.4 µM (Figure 1c). In 1987 $Si(OH)_4$ concentrations were greater than NO_3 whereas the converse was true for the April 1988 data.

During September, within the upwelling season when easterly winds are strong /11/, the greatest east-west variability in nutrient concentrations is seen with low values of both NO_3 and $Si(OH)_4$ to the west, gradually increasing to reach the maximum values east of 110°W (Figure 1b). As observed in the April data, both NO_3 and $Si(OH)_4$ concentrations are lower in 1987 compared to 1988 (Figure 1d). The maximum values in the El Niño (Figure 1b) section were half that of September 1988 (Figure 1d), with NO_3 reaching only 7 µM versus 14 µM, and maximum values for $Si(OH)_4$ of 4.5 µM versus 9 µM. Nitrate concentrations are greater than $Si(OH)_4$ concentrations in both September data-sets (Figures 1b, d), especially to the east of the section with the exception of a region between 180° and 130°W in Figure 1b (week ending 87273).

The values of $NO_3:Si(OH)_4$ for each pixel along the equator range from 0 to 1.7 (Figure 2). In April 1987 (Figure 2a) this ratio was below 1 across the entire longitudinal profile, indicating NO_3 limitation, whereas in April 1988 the ratio was about 1.5 (Figure 2c) over most of the section, with some values <1 to the west. The most east-west variability (i.e. higher concentrations) is shown in the September 1987 section (Figure 2b), with $NO_3:Si(OH)_4$ <1 and declining west of 140°W and >1 east of this longitude. The 1988 September values (Figure 2d) show the same increase to the east but with nearly all values along the equator >1 (i.e. $Si(OH)_4$ in limiting supply), increasing to >1.5 east of 150°W and reaching 1.7.

The north-south variability is best illustrated using the September 1988 image (Figure 3) which had the greatest range in SST and nutrient concentrations of all images used in this study. The nutrient enrichment along the equator due to equatorial upwelling is evident in the eastern equatorial Pacific, particularly for NO_3, such that the equatorial swath shows a $NO_3:Si(OH)_4$ ratio greater than 1.4 from 90°W to 180°W, i.e. along this band $Si(OH)_4$ is most likely more limiting than NO_3 in a Liebig sense /16/. There is a tendency for the NO_3 enhancement to be stronger than $Si(OH)_4$ with maximal concentrations of both tending to occur to the south of the equator, with silicate limitation to the south of the equator. Nitrate limitation (i.e. $NO_3:Si(OH)_4$ <1) tends to occur to the north of the equator, a result of the persistence of $Si(OH)_4$. Because $Si(OH)_4$ does not go to zero, the nutrient variability is apparently driven more by the NO_3 concentrations.

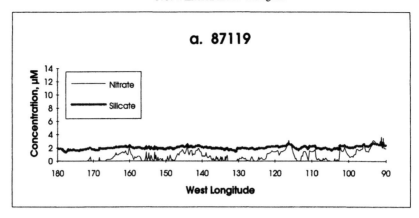

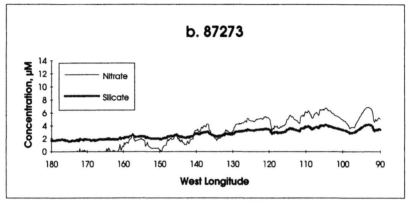

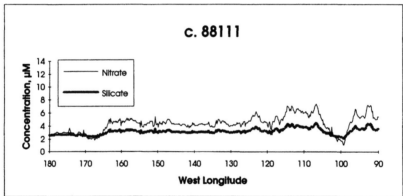

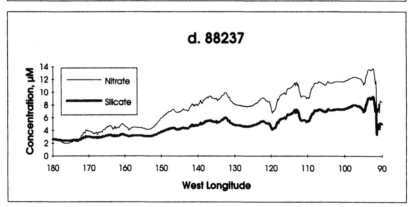

Fig. 1. Equatorial concentrations of surface nitrate and silicate (in µM) from 180°W to 90°W obtained from AVHRR-SST weekly averaged images for weeks ending the Julian Day shown.

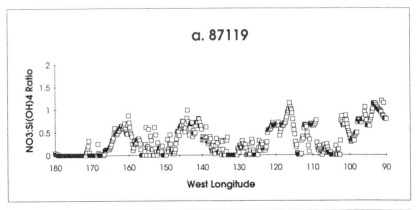

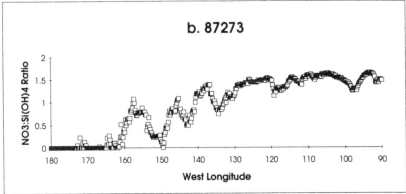

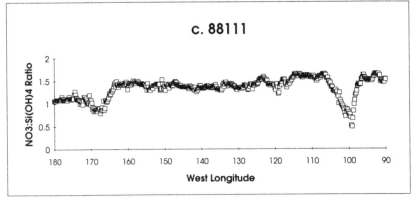

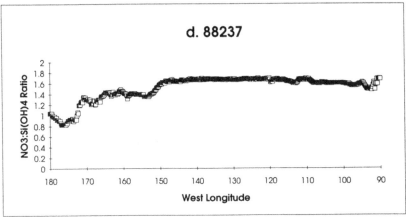

Fig. 2. Equatorial values of the NO_3:$Si(OH)_4$ ratio from 180°W to 90°W obtained from AVHRR-SST weekly averaged images for weeks ending the Julian Day shown.

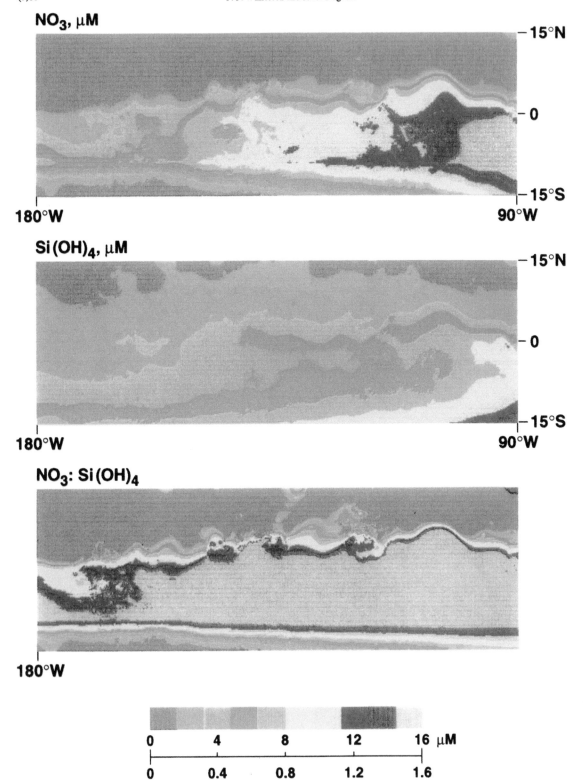

DISCUSSION

These nutrient data for the equatorial Pacific, obtained from AVHRR derived SST data, compare well with shipboard measured values available for this area (e.g. /19/), indicating that this approach can be used to obtain $Si(OH)_4$ in addition to NO_3 concentrations /6/. During the WEC88 study at 150°W in February and March 1988, surface equatorial values of NO_3 = 5 µM and $Si(OH)_4$ = 2.5 µM were measured /20/ with a maximum of 6 µM for NO_3 observed during a time series experiment at the same location /21/. Our data for April 1988 (Figure 1c) show values at 150°W of NO_3 = 4.2 µM and $Si(OH)_4$ = 3 µM. Bender and McPhaden /22/ measured surface values of NO_3 (4 to 5.5 µM) and $Si(OH)_4$ (1 to 2 µM) at the equator at 135°W in April 1988, compared to our estimates of 4.05 and 3.05 µM (Figure 1c). In addition they /22/ reviewed the Hawaii-Tahiti Shuttle nutrient data from March to April 1980 /23/ at 150°W (maximum NO_3 = 6 µM, $Si(OH)_4$ = 3 µM), at 153°W (NO_3 = 4.7 µM, $Si(OH)_4$ = 2.6 µM) and at 158°W (NO_3 = 3 µM, $Si(OH)_4$ = 2.6 µM).

Like the nutrient distributions obtained by us from SST for 1988, the Shuttle data shows the east-west decrease in NO_3, with less variation in $Si(OH)_4$ values. Tilting of the thermocline on a basinwide scale occurs /24, 12/ so that the upper boundary of the thermocline on the equator is shallower to the east, about 40 m at 110°W compared to 100 m at 140°W. As a consequence the east side is usually higher in nutrients than the western side. The south-north changes in NO_3 and $Si(OH)_4$ resemble those described by Carr et al. /20/ for 150°W, with higher NO_3 than $Si(OH)_4$ to the south of the equator and the converse to the north.

As in the 1982-3 El Niño event /25, 26/, nutrient concentrations were reduced during the 1987 El Niño. This was especially noticeable during the upwelling season (Figure 1b). Halpern and Feldman /7/ also noted there was a smaller upward flux of NO_3 in 1987 due to the El Niño conditions, that substantially reduced the surface layer phytoplankton pigments observed by the coastal zone color scanner (CZCS). Interestingly NO_3 shows stronger spatial and temporal variability both intra- and inter-annually than $Si(OH)_4$ (Figure 1). Yang /8/ also noted that $Si(OH)_4$ concentration changes were smaller than NO_3 and that $Si(OH)_4$ did not vary greatly in the equatorial Pacific. This is most likely because the source water for upwelling (the Equatorial Undercurrent) is low in $Si(OH)_4$ compared to NO_3 /8, 19/ setting a limit on its availability and resulting in it being the limiting macronutrient for diatom growth in most situations. In addition, Yang /8/, Wyrtki and Kilonsky /19/ and Bender and McPhaden /22/ showed surface $Si(OH)_4$ never to be completely depleted and reach zero, with the lowest observed values to be about 2 µM whereas NO_3 goes to undetectable levels, as shown also by our values from SST data (Figure 1).

When our approach using AVHRR sea-surface temperatures was used to estimate the concentrations of these nutrients, the ratios of NO_3:$Si(OH)_4$ values gave comparable results to others (e.g. /8/). During warmer years (1987 versus 1988) or seasons (April versus September) the NO_3:$Si(OH)_4$ ratios suggest NO_3 limitation (ratios are less than one), whereas during the cooler years (1988) and seasons (September versus April) $Si(OH)_4$ limitation is more likely. There is a tendency during non El Niño conditions for a NO_3:$Si(OH)_4$ ratio <1 to the west indicating NO_3 to be the limiting nutrient, and so here this may be an HNLC region /27/ whereas to the east the ratio is typically >1, and NO_3 may be high but another nutrient ($Si(OH)_4$) is low, confirming our view that for much of the time this region is actually HNLSLC, in agreement with our examination of other HNLC regions /2/.

In conclusion, we have been able to successfully estimate NO_3, $Si(OH)_4$ and the ratios of NO_3:$Si(OH)_4$ in the eastern equatorial Pacific using remotely-sensed SST data and nutrient-temperature regressions made on shipboard. Silicate concentrations are always above zero, but typically with values less than NO_3 suggesting $Si(OH)_4$ to be in minimal supply. The exception is during warmer situations, i.e. the western part of the equatorial swath or all along the equator during El Niño conditions when NO_3 goes to undetectable levels, and is likely to be the limiting macronutrient.

ACKNOWLEDGEMENTS

Financial support for this research was provided by NASA (161-30-33 to C.O. Davis and 162-MD/BGO-005-91 to RCD and FPW). We wish to thank D. Halpern for obtaining the MCSST data and F. Chavez for use of EPOCS 1988 nutrient data.

REFERENCES

1. H.J. Minas, M. Minas and T.T. Packard, Productivity in upwelling areas deduced from hydrographic and chemical fields, *Limnol. Oceanogr.*, 31, 1182-1206, (1986).

2. R.C. Dugdale, F.P. Wilkerson and H.J.Minas, The role of a silicate pump in driving new production, *Deep-Sea Res.*, in press.

3. A.F. Michaels and M.W. Silver, Primary production, sinking fluxes and the microbial food web. *Deep-Sea Res.*, 35, 473-490, (1988).

4. R. Legeckis, Long waves in the eastern equatorial Pacific Ocean: a view from a geostationary satellite, *Science*, 197, 1181-1197, (1977).

5. D. Halpern, R.A. Knox and D.S. Luther, Observation of 20-day period meridional current oscillations in the upper ocean along the Pacific equator, *J. Phys. Oceanogr.*, 18, 1514-1534, (1988).

6. R.C. Dugdale, F.P. Wilkerson, D. Halpern, F.P. Chavez and R.T. Barber, Remote sensing of seasonal and annual variation of equatorial new production: a model for global estimates, *Adv. Space Res.*, 14, 169-178 (1994).

7. D. Halpern and G.C. Feldman, Annual and interannual variations of phytoplankton pigment concentration and upwelling along the Pacific equator, *J. Geophs. Res.*, 99, 7347-7354, (1994)

8. S.R. Yang, *Factors Controlling New Production in the Equatorial Pacific*, Ph. D. Thesis, University of Southern California, 1992.

9. F.P. Chavez, S. Service and S.E. Buttrey, Temperature-nitrate relationships in the central and eastern tropical Pacific, *Deep-Sea Res.*, in press.

10. D. Halpern, The NASA Ocean Data System at the Jet Propulsion Laboratory, *Adv. Space Res.*, 11, (1991).

11. D. Halpern, Observations of annual and El Niño thermal and flow variations at 0°, 110°W and 0°, 95°W during 1980-1985, *J. Geophys. Res.*, 92, 8197-8212, (1987).

12. G. Philander, *El Niño, La Niña and the Southern Oscillation*, Academic Press, San Diego, 1990.

13. M.J. McPhaden and S.P. Hayes, Variability in the eastern equatorial Pacific during 1986-1988, *J. Geophys. Res.*, 95, 13195-13208, (1990).

14. F.P. Chavez, K.R. Buck and R.T. Barber, Phytoplankton taxa in relation to primary production in the equatorial Pacific, *Deep-Sea Res.*, 37, 1833-1752, (1990).

15. C. McClain, G. Fu, M. Darzi and J. Firestone, *PC-SEAPAK Users Guide*, NASA/GSFC Laboratory for Oceans, 1989.

16. J. Liebig, *Die Chemie in ihrer Anvendung auf Agricultur und Physiologie.* 4th ed. 1847, Taylor and Walton, London, 1840.

17. M.R. Droop, Some thoughts on nutrient limitation in algae, *J. Phycol.,* 9, 264-272, (1973).

18. M. Brzezinski, The Si:C:N ratio of marine diatoms: interspecific variability and the effect of some environmental variables, *J. Phycol.*, 21, 347-357, (1985).

19. K. Wyrtki and B. Kilonsky, Mean water and current structure during the Hawaii-to-Tahiti Shuttle Experiment, *J. Phys. Oceanogr.,* 14, 242-254, (1984).

20. M-E. Carr, N.S. Oakey, B. Jones and M.R. Lewis, Hydrographic patterns and vertical mixing in the equatorial Pacific along 150°W, *J. Geophys. Res.,* 97, 611-626, (1992).

21. F.P. Wilkerson and R.C. Dugdale, Measurements of nitrogen productivity in the equatorial Pacific at 150°W, *J. Geophys. Res.,* 97, 669-679, (1992).

22. M.L. Bender and M.J. McPhaden, Anomalous nutrient distribution in the equatorial Pacific in April 1988: evidence for rapid biological uptake, *Deep-Sea Res.,* 37, 1075-1084, (1990).

23. R.T. Williams, Hawaii-Tahiti Shuttle Experiment Hydrographic Report, Scripps Ref. No. 81-5 to 81-8, (1981).

24. C. Colin, C. Henin, P. Hiscard and C. Oudot, Le Courant de Cromwell dans le Pacifique central en février, *Cah. ORSTOM, Ser. Oceanogr.,* 9, 167-186.

25. R.T. Barber and F.P. Chavez, Ocean variability in relation to living resources during the 1982-1983 El Niño, *Nature,* 319, 279-285, (1986).

26. R.T. Barber and J.E. Kogelschatz, Nutrients and Productivity During the 1982/93 El Niño, in: *Global Ecological Consequences of the 1982-83 El Niño-Southern Oscillation,* ed. P.W. Glynn, Elsevier, Amsterdam 1990, p.21.

27. H.J. Minas and M. Minas, Net community production in "High Nutrient-Low Chlorophyll" waters of the tropical and Antarctic Oceans: grazing versus iron hypothesis, *Oceanologica Acta,* 15, 145-162, (1992).

Adv. Space Res. Vol. 18, No. 7, pp. (7)91–(7)97, 1996
Copyright © 1995 COSPAR
Printed in Great Britain. All rights reserved
0273–1177/96 $9.50 + 0.00

Pergamon

0273–1177(95)00952–3

ESTIMATION OF NEW PRODUCTION FROM REMOTELY-SENSED DATA IN A COASTAL UPWELLING REGIME

R. M. Kudela and R. C. Dugdale

Department of Biological Sciences, University of Southern California, University Park, Los Angeles, CA 90089-0371, U.S.A.

ABSTRACT

We present a comparison of new production estimates derived from traditional shipboard experiments (^{15}N-tracer methodology) with the results of a model incorporating remotely-sensed data from the coastal upwelling region of Monterey Bay, CA (USA) in May, 1993. The model utilized sea-surface temperature derived from the Advanced Very High Resolution Radiometer (AVHRR) to drive a physiologically-based estimate of nitrate uptake in upwelling phytoplankton communities. A direct comparison indicates that the modeled results accurately predict the physiological status of the phytoplankton community. However, these data also demonstrate the difficulty of direct comparison of shipboard estimates of productivity with a synoptically derived estimate for the same study site.

INTRODUCTION

It is well recognized that the oceans are chronically undersampled with regards to biological activity; this problem may be somewhat alleviated with the introduction of a new generation of satellites. There is no electro-magnetic radiation which can be directly correlated to new production (i.e. primary production based on allochthonous sources of nitrogen, primarily NO_3^-). As a result, any modeling efforts require a synthesis of both direct measurements and predictive relationships. Previous remote-sensing models of new production have been based almost exclusively on empirical relationships rather than on mechanistic or physiologically-based relationships, and have been applied to large-scale processes (e.g. /1, 2, 3/). The present study compares new production estimates from traditional shipboard measurements with a model based on the shift-up hypothesis of new production /4, 5, 6, 7/ and utilizing AVHRR imagery. The shift-up hypothesis provides a physiological description of the ecological events occurring within a phytoplankton community as upwelling occurs; as the water mass ages, nitrate utilization is predicted to increase in a non-linear way (acceleration) until a maximum value is reached. As nutrients are depleted, nitrate utilization decreases (shift-down) and regenerated production becomes increasingly important. This model provides both an absolute estimate of new production from satellite imagery and a prediction of the physiological conditions associated with the upwelling phytoplankton community.

METHODOLOGY

Shipboard Experiments

Experiments were conducted in Monterey Bay, CA aboard the R/V *Point Sur* during May 1993 in a grid pattern encompassing *ca.* 500 km^2 over a period of 3 days (Figure 1). At each gridpoint, new production was estimated from ^{15}N-NO_3^- uptake measurements conducted using simulated in situ experiments as described by Dugdale and Wilkerson /8/. Samples for automated nutrient analysis were collected from an instrumented CTD-rosette package, which also provided temperature data. Data from a series of 20-L enclosure experiments conducted in March, May, September and November of 1992 and 1993 as part of the same project were used to provide some of the physiological parameters and physical constants for this model.

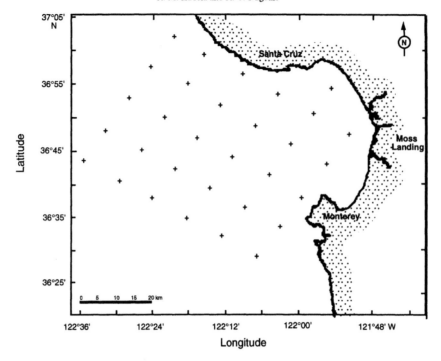

Fig. 1. Map of Monterey Bay, California with gridpoint locations (+). The area mapped corresponds to 65x65 pixel satellite images.

Satellite Data

AVHRR imagery was provided by Ocean Imaging Inc., and consisted of Channel 4 non-atmospherically corrected, 512x512 pixel images, spatially navigated to within one pixel of known coastal points. The images were downloaded from Oregon State University via Internet, to a PC running the Spyglass® (Spyglass Inc.) image processing application where all subsequent processing was completed. All of the images were collected from night-time satellite passes (between 2300 and 0600 PDT) and were cloud-free over the study-site. Each image was calibrated for sea-surface temperature against a minimum of 3 meteorological buoys; criteria for calibration was that the buoy be in a cloud-free area of the image and that buoy data was collected within one hour of the satellite pass. After calibration, the images were spatially reduced by subsampling a 65x65 pixel data-set centered on Monterey Bay (Figure 1), which was used for all subsequent steps. Three images corresponding to Julian Days (JD) 127, 128 and 129, 1993 were combined to form a time-averaged composite by using the mean value for each co-located pixel of the three images. The use of a composite image was chosen to allow comparison of the results of the AVHRR-driven model with the ship-collected data, which was collected during the same time period.

Model Description

The model presented is derived from a similar model originally published by Dugdale et al. /1/; it predicts new production (as NO_3^- uptake) based on the relationship between water temperature and nutrient concentration (e.g. /8/) and the "shift-up" physiological response. Because the purpose of this study is to determine whether such a model is directly comparable to shipboard measurements, we have restricted the results to encompass only sea-surface values at this time rather than complicate the results by incorporating a vertical (depth) component. A flow-chart of the model inputs and assumptions is shown in Figure 2 and is discussed in detail below.

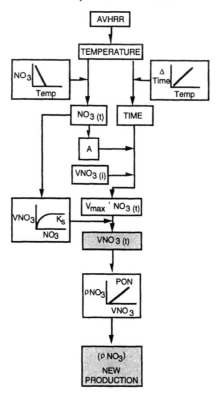

Fig. 2. Schematic diagram of the steps in this model. The gray-shaded boxes indicate the modeled results, specific nitrate uptake rate (V; h^{-1}) and volume transport rate (ρ; mg-at N m^{-3} h^{-1}), equivalent to new production.

Temperature. Surface nitrate concentration and the age of the water for each pixel was estimated from the AVHRR sea-surface temperature values. Nitrate concentrations were estimated using a best-fit linear equation with discrete temperature and $[NO_3^-]$ samples collected from <5 m depth from the shipboard-collected water samples:

$$[NO_3^-] = \text{intercept - slope} * T_{sat} \tag{1}$$

where T_{sat} = the AVHRR-derived sea-surface temperature; intercept = 78.3; slope = -5.89; and r^2 = 0.84. Second- and third-degree polynomial equations were also calculated but were rejected since they did not significantly improve the coefficient of determination. Water age (t) for each pixel was determined by assuming that the upwelling source water was of constant temperature (9.0° C) and that the upwelled water had a constant rate of heat absorption, expressed as $\partial T/\partial t$:

$$t = (T_{sat} - 9.0) / (\partial T/\partial t) \tag{2}$$

where $\partial T/\partial t = 0.5$ C° d^{-1}. Source water temperature was determined for the study site from shipboard measurements and confirmed by satellite imagery. The rate of heat absorption, $\partial T/\partial t$, was chosen based on drifter track studies performed in Monterey Bay.

Kinetic Parameters. The following constants were used in formulating this model:

$$D \text{ (h } d^{-1}) = 12 * (-\cos(JD / 58.09) * \sin(Lat))^3 + 1 \tag{3}$$

$$V_i \text{ (}h^{-1}) = 0.0075 \tag{4}$$

$$K_s \text{ (}\mu M \; NO_3^-) = 1.0 \tag{5}$$

$$A_{max} \ (h^{-2}) = 1.34 \times 10^{-3} \tag{6}$$

$$A \ (h^{-2}) = (1/ \ SQRT(2 * \Pi)) * e^{(-0.35 * t^2)} \tag{7}$$

where D = the latitudinally dependent daylength (expressed as hours of daylight per day); V_i = the initial (freshly upwelled) biomass-specific nitrate uptake rate; K_s = the half-saturation constant for NO_3^- uptake; A_{max} = the (empirically derived) potential maximum acceleration rate; A = the time-dependent function of acceleration (or shift-up and shift-down). A = a mathematical formulation for the shape of a non-symmetric "bell" curve, where the x-axis = time and the y-axis = V; shift-up follows the positively increasing slope of the curve, balanced growth is described by the asymptote, and the negative slope describes shift-down. This function is weighted to allow shift-down to occur more slowly than shift-up (e.g. the curve is asymmetrical). These constants were chosen based on previous large-volume (20-L) simulated in situ uptake experiments conducted under a variety of conditions in Monterey Bay during 1992 and 1993. Using these constants and the data derived from the satellite imagery (Eqns. 1, 2), three equations were formulated which describe biomass-specific nitrate uptake rate:

$$V_{max}' = V_i + (t_{Amax} * D * A_{max}) \tag{8}$$

$$V_{max}(t) = V_{max}' \text{ as a function of } A \tag{9}$$

$$V_{sat} = V_{max}(t) * [NO_3^-] / (K_s + [NO_3^-]) \tag{10}$$

Where V_{max}' = the calculated potential maximal specific nitrate uptake rate assuming complete physiological shift-up (i.e. complete shift-up has occurred); $V_{max}(t)$ = the maximal, non-nutrient limited specific nitrate uptake rate as a function of the shift-up response (i.e. $V_{max}(t) \leq V_{max}'$); and V_{sat} = the satellite-derived prediction of V assuming both shift-up and Michaelis-Menten nutrient kinetics. To directly compare shipboard (24 h incubations) and modeled specific uptake rates (V), all rates were converted to a per-hour time scale.

New Production Estimation. In addition to specific nitrate uptake rates (V), we also compared absolute uptake (transport) rates (ρ; mg-at N m^{-3} h^{-1}). Because PON estimates (which are necessary to calculate a direct estimate of ρ from V) are not available from satellite data, we chose to estimate ρ based on an empirical relationship between V and ρ calculated from shipboard data.

RESULTS

Figure 3 shows the results of the model with the shipboard NO_3^- uptake data superimposed for direct comparison. Surface temperature (Figure 3a) and estimated NO_3^- concentrations (Figure 3b) demonstrate a strong, nutrient replete upwelling plume to the north of the bay off Santa Cruz, with a secondary upwelling plume further south at Pt. Lobos. Temperature rapidly increases and nutrients decrease within the bay itself, while an intrusion of warm water from the California Current system (located west of the study site) is also evident.

NO_3^- uptake rates (V; Figure 3c) and new production (ρ; Figure 3d) demonstrate two bands of relatively high values, one band originating off the coast of Santa Cruz, where localized upwelling was occurring, and the second, offshore band extending from further north, probably from the upwelling feature located off Pt. Reyes (north of our study-site; see /9/ for a discussion). These two prominent bands of high productivity combine further south, where our model predicts a large area of high NO_3^- uptake and new production. This feature is likely the result of the southward advection of the maturing phytoplankton community. Other prominent features include the frontal zone created between the two bands of high productivity, and the intrusion from the west of low productivity, low nutrient water from the California Current system.

As can be seen from the superimposed contours (Figures 3 c, d) the shipboard data and the model are qualitatively in close agreement. To make a quantitative evaluation between predictions from the

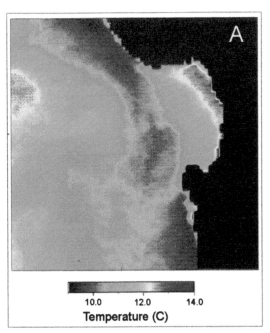

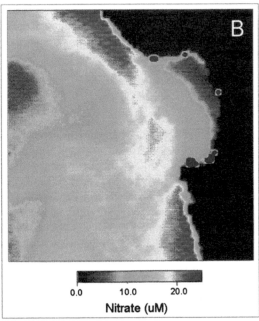

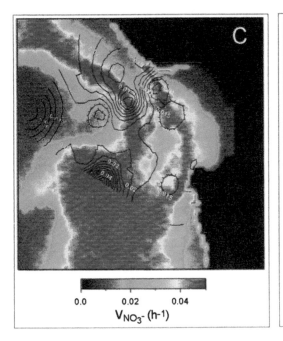

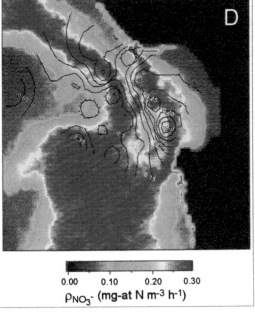

Fig. 3. Model results for Monterey Bay, California calculated using a 3-day composite image (Julian Days 127-129). A: AVHRR-derived temperature (°C), B: modeled [NO3⁻] (µM), C: specific NO3⁻ uptake rate (V, h⁻¹) and D: mass transport rate (ρ, µg-at N m⁻³ h⁻¹) are shown. Panels C and D have superimposed contours of the shipboard experimental results for comparison to the modeled output.

model and the field data, individual model data points were directly compared to the corresponding shipboard experiments. The results of this comparison demonstrate a much better predictive capability for V, or specific uptake of NO_3^-, than for ρ, absolute new production (n = 13; correlation coefficient r of 0.581 and 0.021 respectively). The poor correlation is likely due to the inherent problems in comparing discrete shipboard data points with a temporally composite model, especially in a dynamic environment such as an upwelling plume. Although better correlations are obtained by limiting the comparison to ship incubations conducted during the same 24 h period as individual satellite images (i.e. non-composite images), this reduces the sample size to ≤ 4. Additionally, estimates of absolute new production (ρ) calculated from this model are based on an empirical relationship which has variability and error associated with it.

CONCLUSIONS

Our preliminary efforts to estimate new production from remotely-sensed data demonstrate that it is both possible and desirable to determine both the rates of new production in coastal upwelling regimes, and to predict the physiological condition of the phytoplankton community. We have demonstrated that the AVHRR-derived sea-surface temperature data commonly available may be used successfully to predict surface nitrate concentrations. When these data are coupled with some basic knowledge of the phytoplankton community in the study area, a physiologically-based model of nitrogen utilization may be applied. The results of such a model predict accurate estimates of specific nitrate uptake rates, and allow for an estimation of absolute new production rates.

Although the scope of our model is currently limited to surface estimates of new production, we feel that this data set provides an important first step in the development of models capable of predicting complex ecosystem interactions without resorting to "black box" empirical relationships. This data also demonstrates the inherent difficulty of comparing satellite-derived model predictions with traditional ship-board measurements. Although direct comparisons are difficult, the results of such a study provide a model that accurately represents the biological processes occurring in the dynamic coastal upwelling region of Monterey Bay. Such predictions are not possible with previously published models of new production, which were developed for implementation over much larger geographical areas. We feel that in the future, region-specific models such as this one will provide essential complementary data not readily available from either the spatially limited ship-based field studies or the large-process remote sensing models.

REFERENCES

1. R.C. Dugdale, A. Morel, A. Bricaud and F.P. Wilkerson, Modeling new production in upwelling centers: II A case-study of modeling new production from remotely sensed temperature and color, *J. Geophys. Res.* 94, 18119, (1989).

2. S. Sathyendranath, T. Platt, E.P.W. Horne, W.G. Harrison, O. Ulloa, R. Outerbridge and N. Hoepffner, Estimation of new production on the ocean by compound remote sensing, *Nature*, 353, 129, (1991).

3. P. Morin, M.V.M. Wafar, and P. Le Corre, Estimation of nitrate flux in a tidal front from satellite-derived temperature data, *J. Geophys. Res.*, 98, 4689, (1993).

4. J.J. MacIsaac, R.C. Dugdale, R.T. Barber, D. Blasco, and T.T. Packard, Primary production cycle in an upwelling center, *Deep-Sea Res.*, 32, 503, (1985).

5. ' R.C. Zimmerman, J.N. Kremer and R.C. Dugdale, Acceleration of nutrient uptake by phytoplankton in a coastal upwelling ecosystem: a modelling analysis, *Limnol. Oceanogr.* 32, 359, (1987).

6. R.C. Dugdale and F.P. Wilkerson, New production in the upwelling center at Point Conception, California: temporal and spatial patterns, *Deep-Sea Res.*, 36, 985, (1989).

7. R.M. Kudela, W.P. Cochlan and R.C. Dugdale, Physiological evidence of the shift-up phenomenon in upwelled waters: carbon and nitrogen uptake response by phytoplankton, *Limnol. Oceanogr.*, in press.

8. R.C. Dugdale and F.P. Wilkerson, The use of 15N to measure nitrogen uptake in eutrophic oceans; experimental considerations, *Limnol. Oceanogr.*, 31, 673, (1986).

9. S.-J. Zentara and D. Kamykowski, Latitudinal relationships among temperature and selected plant nutrients along the west coast of North and South America, *J. Mar. Res.*, 321, (1977).

10. L.K. Rosenfeld, F.B. Schwing, N. Garfield and D.E. Tracy, The origin of upwelled water in Monterey Bay: an example of offshore and equatorward advected tongues from an upwelling center. *Cont. Shelf Res.*, 14, 931, (1994).

Pergamon

Adv. Space Res. Vol. 18, No. 7, pp. (7)99–(7)115, 1996
Copyright © 1995 COSPAR
Printed in Great Britain. All rights reserved
0273–1177/96 $9.50 + 0.00

0273–1177(95)00953–1

A LAGRANGIAN MODEL OF PHYTOPLANKTON DYNAMICS IN THE NORTHWEST AFRICAN COASTAL UPWELLING ZONE

A. J. Gabric,* W. Eifler** and W. Schrimpf**

* Faculty of Environmental Sciences, Griffith University, Nathan, Queensland 4111, Australia
** Institute for Remote Sensing Applications, Joint Research Centre, Ispra Site, 21020 Ispra, Italy

ABSTRACT

The growth and cross-shelf advection of phytoplankton in a coastal upwelling zone has been modelled using a time-dependent, two-dimensional Lagrangian-particle approach where the velocity and temperature fields were predicted by a three-dimensional, baroclinic hydrodynamic model. Individual phytoplankton cell parcels are tracked as they move offshore during the development of an upwelling episode. To allow for possible sinking of phytoplankton cells, neutrally buoyant nutrient particles are treated separately in the model. Phytoplankton specific growth is assumed to be limited by a combination of available nutrients, light and temperature.

The Lagrangian method allows the inclusion of the results of recent drogue experiments on the variation of nutrient uptake rate as phytoplankton cells reach the euphotic zone and adjust their physiological status to the high light regime. The model sensitivity to the strength of wind mixing, characteristics of the upwelled source water and initial nutrient and phytoplankton distributions have been investigated. Model predictions suggest nitrate will be depleted seaward from the shelf break and that nutrients will limit phytoplankton growth in offshore waters. Under favourable wind conditions phytoplankton filaments up to 100 km offshore can be formed.

INTRODUCTION

Coastal upwelling zones occur along eastern oceanic boundaries where large-scale weather patterns produce consistent longshore wind stress which induces Ekman transport of surface waters offshore and the vertical advection of cooler, oceanic waters onto the continental shelf. The upwelled waters are usually a rich source of new plant nutrients (NO_3^-) to the coastal ecosystem and promote high primary productivity, a large proportion of which is new production as defined by [1]. It is this new production, as distinct from regenerated production from food-web recycling, which can lead to the ultimate removal of particulate organic carbon (POC) either as a fish yield, burial in the coastal sediments or by export to the deep ocean. These ocean margin processes are of fundamental importance to the understanding of the global carbon cycle [2] albeit as yet poorly analysed. Recent evidence for a climate-change related intensification of upwelling in several different coastal regions [3] suggests increased primary production is likely in these areas, although it is not clear whether this extra production will be utilised by higher trophic levels or indeed how this will effect cycling of shelf organic carbon.

Phytoplankton growth in coastal upwelling zones is greatly influenced by the interaction of physical processes and algal cell physiology. The shelf topography, the time scale and intensity of wind fluctuations, vertical turbulence and heating of the surface layer all have important roles in determining the spatio-temporal characteristics of the phytoplankton distribution. For example, strong vertical mixing leading to a deep mixed layer and high turbidity can limit growth in the Northwest African upwelling region where maximal rates are only achieved during wind relaxation /4/.

Seed algal cells originate from below the euphotic zone and it has been recognised from studies at Peru - 15°S /5/, and Point Conception, California /6/, that a sequence of physiological changes occur as these cells reach the surface layer and move offshore from the upwelling centre. Newly upwelled phytoplankton have low nitrate uptake rates, but as the cells adapt or 'shift-up' their metabolism to the high light intensities in the upper layer, higher nitrate uptake rates are achieved, and later carbon fixation rates also increase. As the cells move further offshore nutrients are depleted and the nitrate uptake rate decreases - a shift-down phase. The timing of maximal growth rate is important, as depending on whether the upwelling centre is located on the inner, mid or outer shelf, it is possible that high rates are achieved in slope or offshore waters. Lampitt /7/ has demonstrated the potential for rapid sinking of algal POC below the thermocline after a surface bloom which would effectively prevent this carbon from being recycled on a time scale of decades.

Zimmerman *et al.* /8/ have used a zero-dimensional simulation model to evaluate the effects of irradiance and initial nitrate concentration on the timing of nutrient utilisation during upwelling. In a comparison of ambient conditions at upwelling centres at Point Conception, Peru and Northwest Africa these authors concluded that the time to exhaust nutrient supply depends on mixed layer depth (MLD) and initial nitrate concentration and varies between 10 days for Peru to over 20 days for Northwest Africa, where mixed layers are deeper and light limitation impedes the attainment of full physiological shift-up. In the Northwest African zone a typical offshore velocity of 20 cms^{-1} in the upper layer would cause a phytoplankton cell to be advected about 17 km per day, and so possibly transported hundreds of kilometres offshore over a 20 day period.

It has been noted /4/ that the offshore-onshore mass balance during upwelling at Cape Blanc is approximately two-dimensional, suggesting that a three-dimensional model is not necessary. Here we present a two-dimensional Lagrangian-particle model of the physiological adaptation and growth of phytoplankton in an upwelling system. Our aim is to investigate the relative impact on phytoplankton growth of the various physical, e.g. wind strength and mixed layer depth, and biological controlling factors such as nitrate concentration, on a time scale of twenty days. The Lagrangian nature of the model allows the inclusion of recent empirical data on physiological shift-up of algal cells which could not be incorporated in a standard Eulerian framework.

Gabric *et al.* /9/ estimated export of shelf production in the vicinity of Cape Blanc [21°N] by analysing satellite imagery of the Northwest African upwelling zone during three distinct upwelling episodes. In each case, the zone of high biomass extended far offshore and comparison of shelf and oceanic pigment concentrations suggested that phytoplankton growth was occurring in the deep ocean 300-400 km from the shelf break, where nutrients would normally be limiting. The Lagrangian model has been applied to the Cape Blanc upwelling region of Northwest Africa with the aim of interpreting the satellite ocean colour data given by Gabric *et al.* /9/ and investigating the conditions under which offshore phytoplankton filaments can be formed.

METHODS

Model Structure

The standard Eulerian approach to the problem /10/ involves the formulation of phytoplankton and nutrient dynamics in terms of coupled partial differential equations, generally written as,

$$\partial P/\partial t \; + \; u\, \partial P/\partial x \; + \; (w+w_s)\, \partial P/\partial z \; - \; K_z\, \partial^2 P/\partial z^2 \qquad = \; S_P \qquad\qquad (1)$$

$$\partial N/\partial t \ + \ u \, \partial N/\partial x \ + \qquad w \, \partial N/\partial z \quad - \quad K_z \, \partial^2 N/\partial z^2 \quad = \quad S_N \qquad (2)$$

with $P(x,z,t)$ the phytoplankton concentration (μg chl a l^{-1}), $N(x,z,t)$ the nitrate concentration (μg-at N l^{-1}), $u(x,z,t)$ and $w(x,z,t)$ the cross-shelf and vertical components of the velocity field and the parameter w_s is the algal cell sinking velocity. K_z is the vertical turbulent diffusion coefficient (m^2s^{-1}) and horizontal diffusion has been neglected. $SP(x,z,t)$ and $S_N(x,z,t)$ are non-linear source terms that describe the phytoplankton and nutrient dynamics, respectively. In the context of an upwelling zone the coordinate system can be chosen such that x denotes the cross-shelf direction (positive offshore) and z the vertical dimension (positive downwards). In the formulation above longshore variability has been neglected. Integration of (1) and (2) requires specification of initial distributions for P and N at time t=0, and definition of boundary conditions at the sea surface, bed and offshore open boundary. If herbivorous zooplankton are to be treated dynamically another similar equation is needed.

In order to investigate the extent of phytoplankton growth offshore of the continental shelf, the model geometry extends 120 km in the offshore direction, about twice the shelf width at Cape Blanc. As we are interested primarily in the photic zone, the vertical dimension of the computational domain is limited to 100m depth.

North-east trade winds predominate in the Cape Blanc region with steady, longshore winds with speeds of 5-10 ms^{-1} and periods of 7-10 days, separated by shorter relaxation periods. The wind stress vector τ (nm^{-2}) at the sea-surface was derived using the conventional quadratic expression,

$$\tau \ = \ \rho_a \ c_d \ |U| \, U$$

where U is the wind-field vector (normally recorded at 10 m), ρ_a is the density of air (1.25 kg m^{-3}) and c_d the drag coefficient (0.0014).

A separate three-dimensional, baroclinic hydrodynamic model /11/ has been used to compute the horizontal and vertical components of the velocity field, $u(x,z,t)$ and $w(x,z,t)$, the turbulent kinetic energy and the temperature field that result from the specification of typical wind stress forcings. The initial temperature distribution was derived from data given by /12/, while the initial salinity distribution was obtained using an empirical relationship with temperature, assuming the water was a mixture of North Atlantic Central Water (NACW) and South Atlantic Central Water (SACW) /13/. Velocities, turbulent kinetic energy and sea-surface anomaly were all set to zero initially. Surface heating was calculated using the radiation model of Brock /14/. In order to avoid a daily net heat flux across the sea surface, the surface boundary condition in the temperature field calculation has been specified as a constant outgoing heat flux equal to the daily average short wave heating.

After an initial numerical 'spin up' period, the hydrodynamic model was run for 20 days and the predicted distributions stored and later used in the biological model simulations. The biological and hydrodynamic models are not coupled in our approach. We have assumed that phytoplankton growth will not significantly affect the thermal structure of the ocean in the upwelling zone. The possible impact of algal shading on the thermal physics of the mixed layer has been discussed by Simonot *et al.* /15/. Mixed layer depth and the strength of vertical mixing due to turbulence, $K_z(x,z,t)$, have been derived from the hydrodynamic model prediction of turbulent kinetic energy.

Rather than attempting to integrate equations (1) and (2) in an Eulerian framework, we have adopted a Lagrangian-particle approach (described below) whereby the mass of phytoplankton and nutrient in each (x,z) grid cell is divided amongst a finite number of P and N particles which interact and are advected by the flow field at each time step.

Rather than starting the simulation with zero values for N and P on the shelf, the values commonly found in upwelled water have been used /16/. The initial N concentration is specified to be constant everywhere on the shelf at a value of 7.4 μg-at N L^{-1}, and the initial P concentration has been set at 0.0 μg chl a L^{-1} above the mixed layer depth and 0.36 μg chl a L^{-1} below the mixed layer (all P particles are in a shifted-down state initially). New N and P particles enter the domain along a vertical boundary at the upper slope (x = 45 km) extending from the sea-bed to the top of the onshore flow, typically 20-30 m. Boundary concentrations are set equal to the initial concentrations given above.

The source terms Sp and S_N specify the growth rate of phytoplankton and depletion rate of nutrient, respectively. Several factors may effect the specific growth rate μ (light, temperature and nutrient limitation), and there are losses L due to respiration and zooplankton grazing. Thus,

$$S_P = \mu (1 - L) P \tag{3}$$
$$S_N = -\gamma \mu (1 - \varepsilon L) P \tag{4}$$

with γ the currency conversion factor between nitrogen and chlorophyll a. The parameter ε is the efficiency by which nutrients are regenerated in the food chain.

Assuming that both availability of light and the ambient temperature can effect the specific nutrient uptake rate at any given time - the multiplicative model /17/, we may write,

$$\mu = V_N R_L R_T \tag{5}$$

where V_N (h^{-1}) is the nitrogen-specific nutrient uptake rate, and R_L and R_T are the dimensionless light and temperature limitation coefficients, respectively.

Following Dugdale *et al.* /18/, we parameterise the change in phytoplankton nitrate uptake as,

$$V_N (t,\tau) = V_N(\tau_0) + (dV_N /dt) \tau \tag{6}$$

where $V_N(\tau_0)$ is the initial low value of uptake at some time τ_0 which is when phytoplankton cells first reach the surface layer at the upwelling centre and $V_N (\tau)$ is the uptake rate after an elapsed time $\tau = t - \tau_0$, during which the cells have been in the high light surface layer. It has been shown /8/ that the acceleration term, dV_N /dt , is related to the nitrate concentration at the upwelling centre (x_0 , z_0) where the cells first reach the surface layer by,

$$dV_N /dt = \alpha N(x_0,z_0,t) + \beta \tag{7}$$

where α and β are empirically derived constants.

Clearly in this prescription $V_N (t,\tau)$ is a Lagrangian and not an Eulerian variable, since it cannot be related to a point in (x,z,t) space, but is rather a property of the individual phytoplankton cells, whose value depends on their residence time τ in the well-lit surface layer. After reaching a peak, the uptake rate will eventually decrease due to nutrient limitation which, using the Michaelis-Menten formulation, can be written as,

$$V_N (\tau) =[V_N (\tau) N(x,z,t)] / [K_s + N(x,z,t)] \tag{8}$$

with K_s (μg-at N l^{-1}) the half-saturation constant.

The light limitation of specific uptake rate is modelled using the exponential formulation which takes photoinhibition into account /19/,

$$R_L = [I(x,z,t)/I_{opt}] \exp(1 - I(x,z,t)/I_{opt}) \tag{9}$$

with $I(x,z,t)$ the ambient light intensity (wm-2), and the parameter I_{opt} is the optimal light intensity that maximises gross photosynthesis. The ambient light intensity at depth z is given by the Beer-Lambert law as,

$$I(x,z,t) = I_0(t) \exp(-k_e(x,t) z) \tag{10}$$

with $I_0(t)$ the surface irradiance ignoring any cross-shelf variation, and $k_e(x,t)$ the average extinction coefficient for downwelling irradiance. Phytoplankton self-shading effects are incorporated through an empirical formula /20/, which relates the average extinction coefficient to the depth-averaged phytoplankton concentration $P(x,t)$ in μg chl L^{-1},

$$k_e(x,t) = 0.04 + 0.0088 P(x,t) + 0.054 [P(x,t)]^{2/3} \tag{11}$$

For the short prediction time scale of upwelling episodes, seasonal changes are ignored, and the diurnal variation in surface irradiance $I_0(t)$, is modelled using the prescription given by Brock /14/. In the Cape Blanc region, field measurements /21/ suggest that the mean extinction coefficient (averaged over the euphotic layer) varies from about 0.5 m^{-1} inshore to 0.2 m^{-1} at the shelf break.

There is a clear effect of temperature on the specific growth rate of phytoplankton and an empirical relationship has been derived /22/,

$$\mu(T) = \mu_0 \exp(0.063 T) \tag{13}$$

where μ_0 is 0.035 (doublings h^{-1}) and T the temperature in degrees Celsius. The non-dimensional temperature limitation coefficient R_T can be derived from (13) by noting that the maximum temperature observed during upwelling on the shelf off Cape Blanc is about 21°C, /23/ , thus (13) may be normalised by the maximum growth rate, $\mu_{max} = 0.13$, to give,

$$R_T = \mu_0 \exp(0.063 T)/\mu_{max} \tag{14}$$

For a temperature of 17°C, which is commonly observed on the shelf during upwelling /24/, R_T is approximately 0.8.

The Lagrangian-Particle Method

In order to incorporate the nutrient uptake behaviour described in (6,7) we have adopted a Lagrangian-particle approach broadly similar in conception to that of Woods and Onken /25/. To allow for negative buoyancy of algal cells we employ both phytoplankton and nutrient computational particles. The concentration fields $P(x,z,t)$ and $N(x,z,t)$ may be retrieved at any time step by summing over the mass of all the P and N particles in each grid cell.

The number of particles in a grid cell will affect the accuracy of the concentration estimate. Ozmidov /26/ has shown that this accuracy is proportional to $n / \sqrt{n}$, where n is the number of particles. We have chosen an initial particle density of 100 particles per cell as a compromise between accuracy and computational efficiency. The computational domain has been divided into 40 cells in the horizontal and 20 cells in the vertical which implies the maximum number of particles in the domain is 160000.

The phytoplankton particles may be thought of as representing a water parcel containing a group of algal cells with similar developmental histories. It is then possible to follow the physiological adaptation of a group of similarly positioned cells as they are vertically advected into the euphotic zone and drift seaward across the shelf. The growth of a phytoplankton particle is thus related to both the environmental conditions at the location of the particle and its individual travel history.

The computational phytoplankton particles each have three properties. Thus the jth phytoplankton particle, P_j, is described by, (i) cross-shelf and vertical position (X_j, Z_j), (ii) mass M_j, and (iii) an age or residence time in the euphotic zone, T_j. The computational nutrient particles, N_j, each have two properties, position and mass.

Advection by the velocity field is modelled by updating the position of P and N particles at each time step Δt. Thus, at the nth time step,

$$(X_j)^n = (X_j)^{n-1} + u(X_j, Z_j)\Delta t \tag{15}$$
$$(Z_j)^n = (Z_j)^{n-1} + [w(X_j, Z_j) + w_s + w']\Delta t \tag{16}$$

where $u(X_j, Z_j)$ is the cross-shelf velocity and $w(X_j, Z_j)$ is the vertical velocity at the position of the P or N particle derived by interpolation from the predicted velocity field at grid points neighbouring (X_j, Z_j). To account for the effects of turbulent mixing a random vertical displacement $w'\Delta t$ is added to the vertical advection. The turbulent velocity w' is calculated from the turbulent kinetic energy $E_k(X_j, Z_j)$ at the particle's location as predicted by the hydrodynamic model, by

$$w' = \sqrt{(2E_k / 3)} \tag{17}$$

The sign of w' is determined by a random number generator. In order to ensure that turbulent mixing remains a subgrid scale effect, the advective time step must satisfy,

$$\Delta t \leq \Delta z / w'_m \tag{18}$$

where w'_m is the maximum turbulent velocity, and Δz is the vertical grid size. For a vertical grid size of 5m and typical maximum turbulent velocity of 5 cm/s the advective time step is 100 seconds.

By contrast, the biological processes are much slower and the mass of phytoplankton and nutrient computational particles need only be updated every hour (the biological timestep, Δt_b). For accuracy the flux of new particles into the domain along the open boundary is also updated every hour. The number of new particles entering the computational domain in any time step is computed from the mass flux at that depth. The mass of each new N or P particle is calculated by dividing the total boundary cell mass by 100 and adjusting for the round-off error introduced by having an integral number of particles.

Growth and nutrient depletion are modelled by updating the mass of P and N computational particles at each time step Δt_b. Thus, at the nth biological time step, the mass of the N and P computational particles are,

$$\text{P:} \quad M_j^n = M_j^{n-1} + M_j^{n-1}\mu(1 - L)\Delta t_b \tag{19}$$
$$\text{N:} \quad M_j^n = M_j^{n-1} - M_j^{n-1}\gamma\mu(1 - \varepsilon L)\Delta t_b \tag{20}$$

Model Parameter Estimation

The parameter values assumed for the model are given in Table 1. Where possible, values appropriate for the Northwest African upwelling zone have been used. Parameters such as the half-saturation constant K_s are not readily measurable in upwelling areas and the value used here is typical of those quoted in the literature /27/.

The uptake acceleration constants α and β were derived from experiments using water sampled at Point Conception, California and the initial nitrate uptake rate $V_N(\tau_O)$, was given a value typical of minimum values at upwelling centres /6/. While it is well known that C:Chl ratios can vary over a large range, the currency conversion factor γ (N:Chl) has been derived by assuming a C:N ratio of 6 and C:Chl of 50 /8/. The value of I_{opt}, the optimal light intensity that maximises gross photosynthesis, is that given by Howe /28/ for the Cape Blanc region.

TABLE 1. Model Parameters

Parameter	Value	Units	Definition
α	4×10^{-5}	h^{-2} ug-at N^{-1}	uptake acceleration constant
β	4×10^{-5}	h^{-2}	"
ε	0.5	-	nutrient regeneration efficiency
γ	8.3	μg N $(\mu g$ Chl$)^{-1}$	N : Chl a ratio
I_{opt}	125.0	w m^{-2}	optimal light intensity
K_s	1.0	μg-at N L^{-1}	nitrate half-saturation constant
L	0.1	-	respiration/grazing loss
$V_N(\tau_O)$	5×10^{-3}	h^{-1}	shifted-down uptake rate
w_s	$0.0 - 1.0 \times 10^{-4}$	m s^{-1}	algal sinking velocity

RESULTS AND DISCUSSION

The model has been used to simulate phytoplankton growth and offshore advection over a 20 day period for various wind forcing scenarios. Phytoplankton growth in the surface layer will be dependent on the strength of wind forcing which determines the MLD and hence the average light and nutrient environment. A large MLD can impede the attainment of maximum productivity as the euphotic zone depth Z_e is typically 22 m in the Cape Blanc region /21/, while the MLD can be over 50m for strong mixing.

Another important effect for the biological response, is the complex vertical circulation predicted by the hydrodynamic model, which suggests a two-celled pattern as schematically depicted in Figure 2. Upwelling is predicted up to mid-shelf with downwelling, albeit slightly weaker, in the outer shelf and upper slope. The existence of a density front and downwelling at the shelf edge has been noted in a number of field studies at Cape Blanc , e.g. /29/ and /30/. Clearly this circulation pattern allows for the possibility of some recirculation of N and P particles.

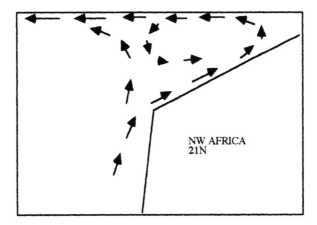

NW AFRICA
21N

Fig. 1. Schematic of two-celled cross-shelf circulation pattern

Wind speed also determines the mean velocity in the surface layer and the overall depth of the offshore flowing layer. Hence the total biomass exported from the shelf and the offshore length scale of the resulting phytoplankton distribution will also be wind speed dependent. Importantly for nutrient uptake, the shifting up process depends on residence time in the surface layer and the location of maximal uptake will thus vary with the strength of offshore advection.

Three steady wind stress cases were simulated: 0.05, 0.10 and 0.15 nm^{-2} corresponding approximately to wind speeds of 5, 7.5 and 10 ms^{-1}, respectively. We have also considered an intermittent wind case where the stress varies from a maximum 0.15 nm^{-2} for 4 days then decreases to 0.01 nm^{-2} for 4 days and repeats on a 10 day cycle as shown in Figure 2.

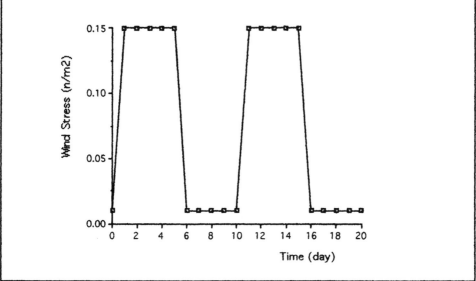

Fig. 2 Intermittent wind stress forcing

Case 1 : Steady Wind : $\tau = 0.15\,nm^{-2}$

The MLD varies diurnally due to the change in solar radiation, with convection due to surface cooling during the night significantly deepening the layer. The MLD ranges from a daytime minimum of about 40 m to a nocturnal maximum of 65m.

The simulated N and P distributions are shown in Figure 3a and 3b at two day intervals. The initial condition of zero P in the mixed layer causes a time lag in N consumption which does not become severely depleted until P fill the entire surface layer at t = 96h. At t = 144 h a patch of nutrient depleted water has been advected to the offshore region (x > 45 km). This uptake is clearly occurring on the shelf as P have not yet been advected offshore at t = 144 h. Thereafter, P begin to move offshore and presumably the combination of *in situ* and previous shelf uptake combine to deplete N up to a depth of 40m in offshore waters. The phytoplankton extend to a distance of about 100 km offshore at t = 480 h. Highest surface layer concentrations (P ≈ 0.5 μg Chl a L^{-1}) occur in the vicinity of the shelf break with concentrations still about 0.2 μg Chl a L^{-1} offshore.

Case 2 : Steady Wind: $\tau = 0.10\,nm^{-2}$

The reduced wind speed results in a daytime minimum MLD of 25m and nocturnal maximum of 60m. The strength of vertical mixing is also reduced compared to Case 1 so that any baroclinic motions near the shelf edge are relatively more effective than in Case 1.

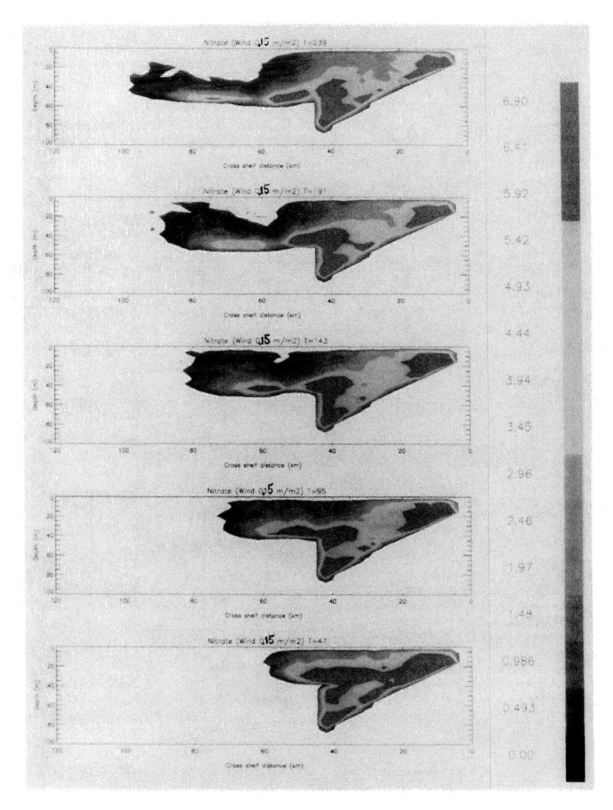

Fig. 3(a) Predicted nitrate (μg-at N L^{-1}) distributions for 0.15 nm^{-2} wind forcing.

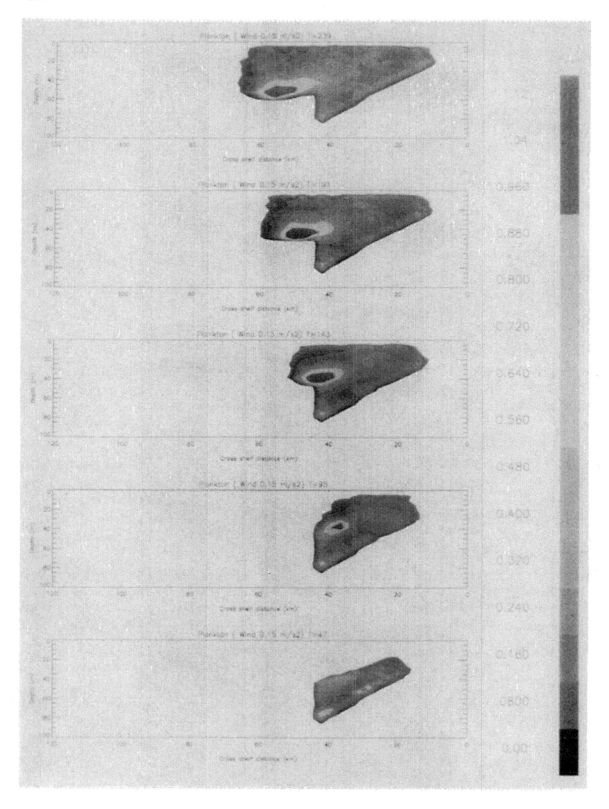

Fig. 3(b) Predicted phytoplankton (μg Chl L^{-1}) time evolution for 0.15 nm^{-2} wind forcing.

The N and P distributions are shown in Figure 4a and 4b. It is interesting that surface N depletion is again noticeable at t = 144 h, however the low nitrate water extends onto the shelf. With the decreased offshore advection in this case, P particles will have longer residence times in the surface layer at the same relative offshore position, and thus attain maximal uptake rates (shift up) closer inshore than in Case 1. While nutrient uptake is higher on the shelf, the chlorophyll values attained in the surface layer on the shelf are similar to those in Case 1. The P distribution extends 75 km offshore at t = 480h.

Case 3: Steady Wind: $\tau = 0.05 \, nm^{-2}$

In this case upwelling is marginal with vertical advective velocities low and confined to the inshore region of the shelf. The reduced wind speed results in a daytime minimum MLD of about 10m and nocturnal maximum of 35m. Vertical mixing strength is about 25% of that in Case 1 and, combined with the very shallow mixed layer, vertical entrainment of phytoplankton into the surface layer is suppressed. Model predictions suggest the P distribution is still confined to the shelf at t = 480 h with the result that nitrate is depleted above 40 m depth.

Case 4: Intermittent Wind

In order to investigate a more realistic wind scenario, intermittency was introduced by applying a wind stress of 0.15 nm^{-2} for 4 days followed by a decrease over one day to a relaxation value of 0.01 nm^{-2} which continued for 4 days . The entire cycle was then repeated. During the cycle the daytime MLD varies between 40m during high wind to less than 5m during relaxation.

N and P distributions are shown in Figures 5a and 5b. It is interesting to compare the N distribution with that of the constant wind stress of 0.15 nm^{-2} shown in Figure 3a. At t = 240h nitrate depletion on the shelf is much higher in the intermittent case with low concentrations from midshelf seaward in the upper 30m of the water column as the upwelling of new nutrients ceases during wind relaxation. A similar situation holds at t = 480h at the end of the next relaxation period. If this type of intermittency is repeated the nitrate concentration on the shelf will be periodically depleted and replenished. The P distribution is confined to the shelf at t = 240h and higher concentrations up to 0.8 µg Chl a L^{-1} are reached, probably due to the reduced vertical mixing during wind relaxation. The offshore extent of the P distribution is around 65 km after 20 days.

Case 5 : High Nitrate Steady Wind : $\tau = 0.15 \, nm^{-2}$

In this simulation the initial and boundary nitrate concentration was doubled to investigate the effects on uptake rate and phytoplankton growth. With a higher ambient N concentration the acceleration term given by (7) will be higher (by a factor 1.8) and maximal uptake rate should be achieved earlier. The model suggests that P concentrations are slightly higher (10%) but generally the distribution is quite similar to Case 1. The N distribution seems to be depleted more at short simulation times (e.g. t = 96h) with values seaward of the 50 km point lower than in Case 1 suggesting higher uptake rates have indeed been achieved.

CONCLUSIONS

The particle model presented here provides a framework for incorporating the physiological hypothesis of shift-up which has recently been suggested by several field investigations in upwelling areas. The Lagrangian nature of this field data cannot easily be incorporated in a standard Eulerian modelling approach.

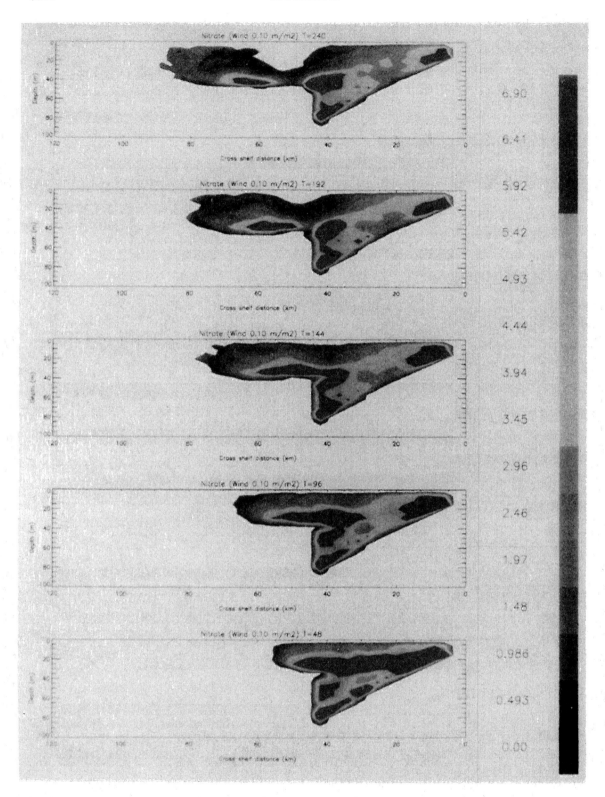

Fig. 4(a) Predicted nitrate (μg-at N L^{-1}) time evolution for 0.10 nm^{-2} wind forcing

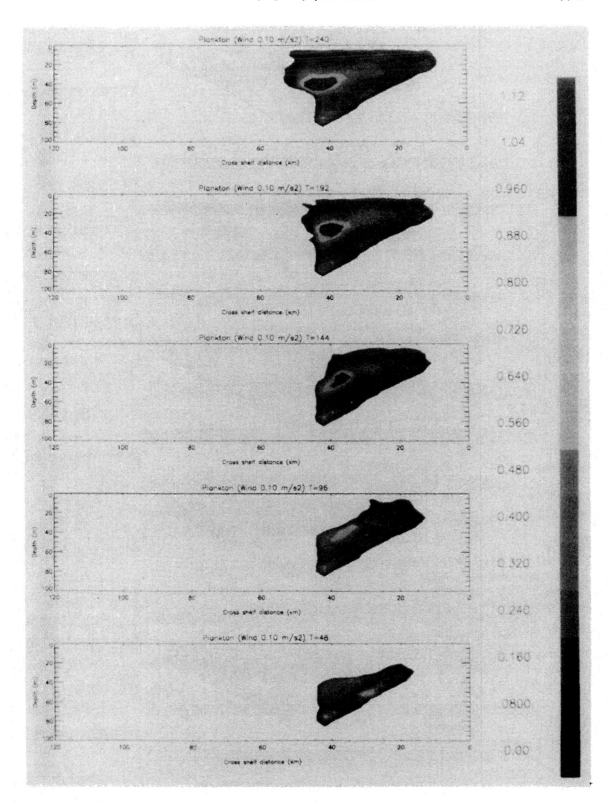

Fig. 4(b) Predicted phytoplankton (μg Chl L^{-1}) time evolution for 0.10 nm^{-2} wind forcing.

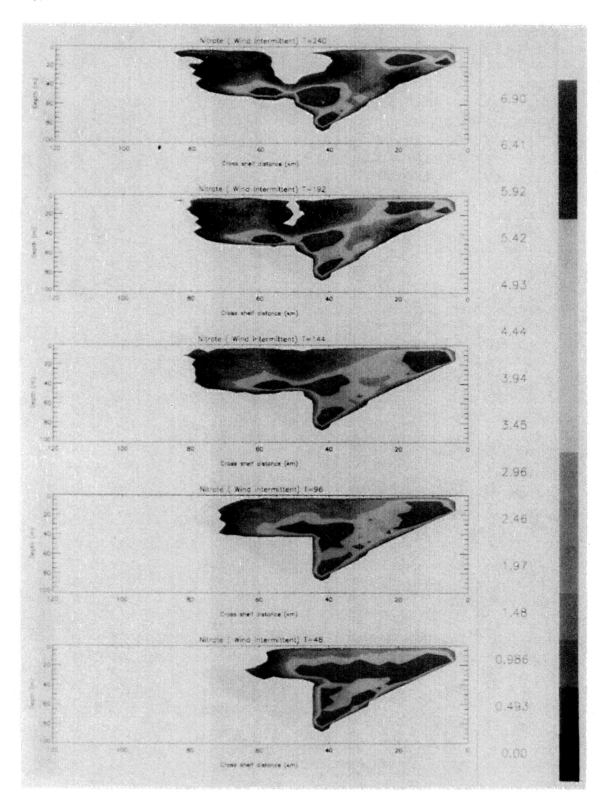

Fig. 5(a) Predicted nitrate (μg-at N L^{-1}) time evolution for intermittent wind forcing.

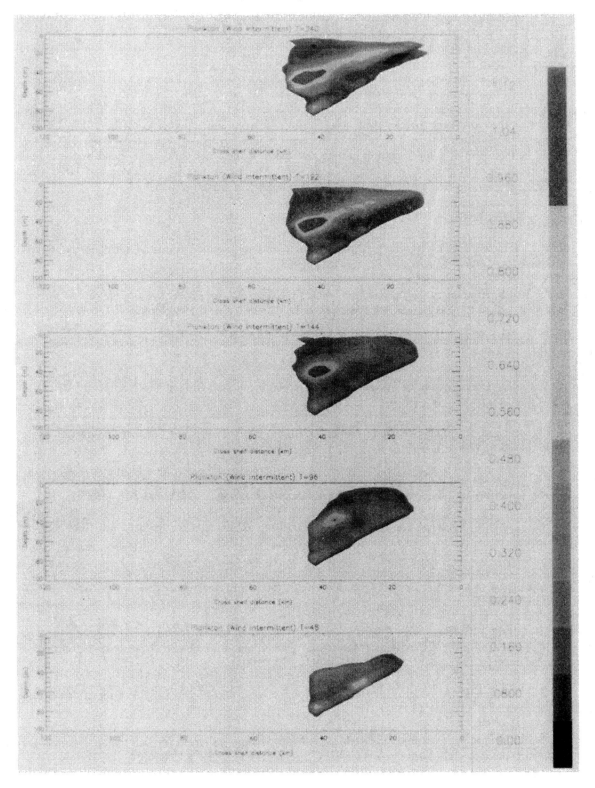

Fig. 5(b) Predicted phytoplankton (μg Chl L^{-1}) time evolution for intermittent wind forcing.

Model predictions suggest algal cells will take up almost all the available nitrate in surface waters seaward of the shelf break with nutrient limitation of growth rate then likely. Importantly however, the phytoplankton plume can still extend up to 100 km offshore after 20 days of constant strong winds, although the more realistic intermittent wind scenario suggests an offshore extent of about 70 km. The model predicts surface nitrate depletion seaward of 30 km in the intermittent case, suggesting that nutrient limitation could exist on the outer shelf. Turbidity due to bottom resuspension of inorganic matter has been neglected here but could reduce growth rate and nitrate uptake on the Northwest African shelf, resulting in a delayed shift-up and an offshore displacement of the phytoplankton plume. The offshore filaments which have been registered by the CZCS, /9/, could be the result of a superposition in time and space of the algal plumes produced by the almost continuous upwelling along the Mauritanian shelf. Mesoscale mixing processes and large scale surface circulation patterns will also affect the morphology of offshore filaments.

REFERENCES

1. R.C. Dugdale and J.J. Goering, Uptake of new and regenerated forms of nitrogen in primary productivity, *Limnol. Oceanogr.*, 12, 196-206, (1967).

2. J.J. Walsh, How much shelf production reaches the deep sea, in, *Productivity of the ocean: present and past,* edited by WH Berger, VS Smetacek and G Wefer, Wiley, 1989, pp175-191.

3. A. Bakun , Global climate change and intensification of coastal upwelling, *Science,* 247, 198-201, (1990).

4. A. Huyer, A comparison of upwelling events in two locations : Oregon and Northwest Africa, *J. Mar. Res.,* 34(4), 531-547, (1976).

5. J.J. MacIsaac, R.C. Dugdale , R.T. Barber, D. Blasco and T.T. Packard, Primary production cycle in an upwelling center. *Deep Sea Res.,* 32, 5, 503-529, (1985).

6. F.P. Wilkerson and R.C. Dugdale , The use of shipboard barrels and drifters to study the effects of coastal upwelling on phytoplankton dynamics, *Limnol. Oceanogr.,* 32(2), 368-382, (1987).

7. R.S. Lampitt, Evidence for the seasonal deposition of detritus to the deep-sea floor and its subsequent resuspension, *Deep-Sea Res.,* 32(8), 885-897, (1985).

8. R.C. Zimmerman, J.N. Kremer and R.C. Dugdale, Acceleration of nutrient uptake by phytoplankton in a coastal upwelling ecosystem: A modelling analysis. *Limnol. Oceanogr.,* 32(2), 359-367, (1987).

9. A.J. Gabric, L.Garcia, L.Van Camp, L. Nykjaer, W. Eifler, and W. Schrimpf, Offshore export of shelf production in the Cape Blanc (Mauritania) giant filament as derived from coastal zone color scanner imagery, *J. Geophys. Res.,* 98, C3, 4697-4712, (1993).

10. J.S. Wroblewski, A model of phytoplankton plume formation during variable Oregon upwelling, *J. Mar. Res.,* 35, 357-394, (1977).

11. W. Eifler and W. Schrimpf, ISPRAMIX, a hydrodynamic program for computing regional sea circulation patterns and transfer processes, CEC Technical Report EUR 14856EN, (1992).

12. H.J. Minas, L.A. Codispoti and R.C. Dugdale, Nutrients and primary production in the upwelling region off Northwest Africa, *Rapp. P-v. Reun. Cons. int. Explor. Mer.,* 180, 148-183, (1982).

13. M. Manriquez and F. Fraga, The distribution of water masses in the upwelling region off Northwest Africa in November, *Rapp. P-v. Reun. Cons. int. Explor. Mer.*, 180, pp. 39-47, (1982).

14. T.D. Brock, Calculating solar radiation for ecological studies, *Ecol. Modelling*, 14, 1-19, (1981).

15. J-Y. Simonot, E. Dollinger and H. Le Treut, Thermodynamic-biological-optical coupling in the oceanic mixed layer, *J. Geophys. Res.*, 93 , C7, 8193-8202, (1988).

16. G. Hempel (ed.), The Canary Current : Studies of an upwelling system, *Rapp. P-v. Reun. Cons. int. Explor. Mer.*, 180, 455pp, (1982).

17. T. Platt, K.L. Denman and A.D. Jassby, Modeling the productivity of phytoplankton, in: *The Sea, Vol.6*, ed. E.D.Goldberg, New York, Wiley, 1977, pp807-856.

18. R.C. Dugdale, F.P. Wilkerson and A. Morel, Realization of new production in coastal upwelling areas: A means to compare relative performance, *Limnol. Oceanogr.*, 35(4), 822-829, (1990).

19. J.H. Steele, Environmental control of photosynthesis in the sea., *Limnol. Oceanogr.*, 7, 137-150, (1962).

20. G.A. Riley, Transperancy-chlorophyll relations, *Limnol. Oceangr.*, 20, 150-152, (1975).

21. R.C. Dugdale, A. Morel, A. Bricaud and F.P. Wilkerson, Modeling new production in upwelling centers: A case study of modeling new production from remotely sensed temperature and color, *J. Geophys. Res.*, 94, C12, 18119-18132, (1989).

22. R.W. Eppley, Temperature and phytoplankton growth in the sea. *Fish. Bull.*, 70, 1063-1085, (1972).

23. E. Mittelstaedt, The ocean boundary along the northwest African coast: Circulation and oceanographic properties at the sea surface, *Prog. Oceanog.*, 26, 307-355, (1991).

24. L. Van Camp, L Nykjaer, E Mittelstaedt and P Schlittenhardt, Upwelling and boundary circulation off Northwest Africa as depicted by infrared and visible satellite observations, *Prog. Oceanog.*, 26, 357-402, (1991).

25. J.D. Woods and R. Onken, Diurnal variation and primary production in the ocean - preliminary results of a Lagrangian ensemble model. *J. Plankt. Res.*, 4, 735-756, (1982).

26. R.V. Ozmidov, *Diffusion of contaminants in the ocean*, Kluwer Academic, Dordrecht, 1990.

27. J.J. MacIsaac and R.C.Dugdale, The kinetics of nitrate and ammonia uptake by natural populations of marine phytoplankton, *Deep-Sea Res.*, 16, 45-57, (1969).

28. S. Howe, A simulation study of biological responses to environmental changes associated with coastal upwelling off Northwest Africa, *Rapp. P V. Reun. Cons. Int. Explor. Mer.*, 180, 135-147, (1982).

29. E. Hagen, Mesoscale upwelling variations off the West African coast, in: *Coastal Upwelling, Coastal Estuarine Sci. Ser., Vol. 1*, ed. F.A. Richards, AGU, Washington, D.C., 1981, pp. 72-78.

30. E. Mittelstaedt and I. Hamann, The coastal circulation off Mauritania. *Dtsch. Hydrogr. Z.*, 34, 81-118, (1981).

Pergamon

Adv. Space Res. Vol. 18, No. 7, pp. (7)117–(7)128, 1996
Copyright © 1995 COSPAR
Printed in Great Britain. All rights reserved
0273–1177/96 $9.50 + 0.00

0273–1177(95)00954–X

USE OF REMOTE SENSING AND MATHEMATICAL MODELLING TO PREDICT THE FLUX OF DIMETHYLSULFIDE TO THE ATMOSPHERE IN THE SOUTHERN OCEAN

A. J. Gabric,* G. Ayers,** C. N. Murray*** and J. Parslow†

* *Faculty of Environmental Sciences, Griffith University, Nathan, Queensland 4111, Australia*
** *CSIRO Division of Atmospheric Research, Private Bag 1, Mordialloc 3195, Australia*
*** *Institute for Remote Sensing Applications, CEC Joint Research Centre, Ispra (Va), I-21020, Italy*
† *CSIRO Division of Fisheries, PO Box 1538, Hobart 7001, Australia*

ABSTRACT

An existing ecological model of DMS production has been extended and applied to the spring-summer period in the Subantarctic Southern Ocean. The model predicts that production of phytoplankton and dissolved DMS will increase during spring to reach a maximum in summer consistent with the atmospheric data collected at the Cape Grim baseline station. Archival Coastal Zone Color Scanner satellite imagery has been used to define the seasonal range in phytoplankton concentration in the study region and validate model predictions . Local measured wind and sea temperature data have been used to calculate the DMS transfer velocity which is used to compute the sea-to-air flux of DMS. The seasonal trend in predicted DMS flux is in good agreement with the flux estimates made from observations.

INTRODUCTION

Dimethylsulfide (DMS; CH_3SCH_3) is an important sulfur-containing trace gas produced by some classes of marine phytoplankton /1; 2; 3; 4 /. DMS is present in oceanic surface waters at concentrations sufficient to sustain a considerable net flux to the atmosphere, which is currently estimated to be 0.5 ± 0.3 Tmol S yr^{-1} compared with the global natural (marine + terrestrial + volcanic) flux estimate of sulfur to the atmosphere of 0.78 Tmol S yr^{-1} /5/.

Charlson et al. /6/ have suggested that a major source of cloud condensation nuclei (CCN) over the oceans is the DMS produced by planktonic algae in seawater. DMS is oxidized in the atmosphere to form non-sea-salt sulfate (nss SO_4^{2-}) and methanesulfonate (MSA) aerosols. Because the formation of clouds is sensitive to CCN density, it has been postulated that biological regulation of the climate is possible by affecting albedo and thus the Earth's radiation budget. This would be due to the effect of temperature and solar radiation on phytoplankton growth; however, it is still unclear whether the change in albedo will cause a positive or negative feedback on climate.

The Southern Ocean is relatively unpolluted and thus the production of sulfate aerosols will be mainly due to the biogenic source of DMS. Measurements made at Cape Grim, Tasmania (40°41'S, 144° 41'E) by Ayers et al. /7/ over a twenty month period have confirmed the connection between atmospheric DMS and aerosol sulfur species - a significant part of the Charlson et al. hypothesis. Sampling at Cape Grim was carried out at 94 m above sea level, only during 'baseline' conditions - periods during which the 10m wind direction is from the southwest quadrant (190° - 280°). A pronounced seasonal cycle in atmospheric DMS was detected with an amplitude of about 10-20. A similar cycle in MSA and n.s.s sulfate was also evident suggesting a tight coupling between DMS, MSA and n.s.s. sulfate aerosols.

While there are very limited measurements of aqueous DMS in the ocean southwest of Cape Grim, northern hemisphere data on oceanic DMS concentrations / 8; 2; 9; 10/ show that average surface seawater DMS concentration may display a seasonal variation of up to a factor of 50 in mid and high latitudes, from a typical winter value of 0.2 nM to a summer maximum of 10 nM. Measurements made by McTaggart and Burton /11/ in the Southern Ocean among 40-53°S during the 1988-89 austral summer recorded mean January DMS of 20.5 nM, a very high value compared to the global mean concentration of about 3nM /12/ or the value of 6.8 nM found in the North Sea during the summer of 1985 /2/.

Here we present a modelling study of the oceanic production of DMS and its flux to the atmosphere under conditions typical of the Subantarctic Southern Ocean southwest of Cape Grim, with the aim of understanding the factors responsible for the amplitude of the measured seasonal cycle in atmospheric DMS. The model of DMS production developed by Gabric et al. /13/, henceforth referred to as the GMSK model, has been extended to include light and temperature effects on phytoplankton growth and the sea-air flux has been formulated in terms of wind speed and sea surface temperature.

MODEL STRUCTURE AND ASSUMPTIONS

The structure of the GMSK model is given in Figure 1 with arrows representing fluxes of nitrogen and sulfur through the food web. The biotic compartments comprise a generic autotroph (phytoplankton), planktonic bacteria which metabolise DMS and its precursor dimethylsulfonium propionate (DMSP), and three heterotrophs. For simplicity, no higher trophic levels are considered although zooplankton export through grazing by fish is included. The model state variables are vertically averaged over the mixed layer depth. The model's ecological structure reflects the current thinking on the role of microorganisms in elemental recycling /14;15 / and the important part played by aerobic bacteria in DMS turnover in the water column /16 /.

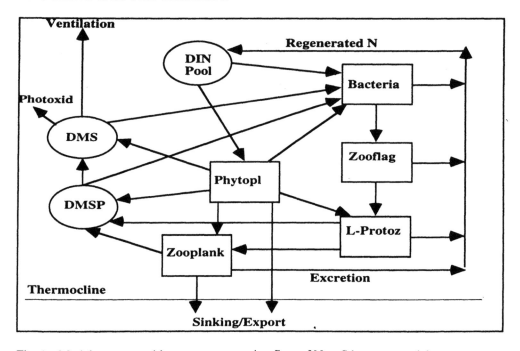

Fig. 1. Model structure with arrows representing flow of N or S between model compartments.

In the extension of the model described here, phytoplankton growth rate is assumed to be affected by light, temperature and nutrient availability. Diurnal variability in the light field is also included. Loss of particulate organic matter (both N and S) occurs by sedimentation below the thermocline. Nitrogen is regenerated via the microbial loop, however to limit the number of rate parameters in the model we have not distinguished between nitrate and ammonium in the nutrient uptake formulation.

Loss of dissolved DMS occurs by photo-oxidation /17/ and ventilation to the atmosphere. Particulate DMS/DMSP can also be lost by sinking of fecal material or dead cells below the mixed layer. Biomass units are all mg N m^{-2}, with the sulfur species measured in mg S m^{-2} . The specific form of the fluxes and parameter values is given in the Appendix with further details available in /13/.

Study Window

The region of the Southern Ocean in which DMS production will affect the measurements of atmospheric DMS made at Cape Grim depends on wind strength, and the lifetime of gaseous DMS in the marine boundary layer. Data are collected only when the wind direction is from 180°-290°, thus avoiding potentially polluted continental air masses.

Wind speeds vary from a monthly mean value of 41 km h^{-1} in January to 26 km h^{-1} in June, so that an air parcel may travel up to 1000 km per day. Since the atmospheric lifetime of DMS varies from approximately half a day in summer to about a week in winter, the monthly mean DMS concentrations observed at Cape Grim may reflect oceanic production in a quadrant well over 1000 km west and south of the sampling station.

Calculation of the Transfer Velocity

In order to adapt the GMSK model to conditions pertaining in the Southern Ocean, the air-sea exchange coefficient or transfer velocity (a constant the original version of the GMSK model) has been parameterised in terms of wind speed and ocean temperature. The wind speed (w) dependence of the transfer velocity k_w (cm h^{-1}) was calculated from the equations given by Liss and Merlivat /18/ which were derived for CO_2 at 20°C and a Schmidt number (Sc) of 600,

$$k_w = 0.17w \qquad\qquad \text{for } w \leq 3.6 \qquad\qquad (1a)$$
$$k_w = 2.85w - 9.65 \qquad \text{for } 3.6 < w \leq 13 \qquad (1b)$$
$$k_w = 5.9w - 49.3 \qquad\quad \text{for } w > 13 \qquad\qquad (1c)$$

These equations must be rescaled to apply for DMS and different Sc. Liss and Merlivat /18/ assume that k_w is proportional to $Sc^{-2/3}$, for wind speeds less than 3.6 ms^{-1} , and $Sc^{-1/2}$ for higher wind speeds. Rescaling equations (1) and enforcing piecewise continuity at the wind speed boundaries gives,

$$k_w = \alpha\, 0.17w \qquad\qquad\qquad\qquad \text{for } w \leq 3.6 \qquad\qquad (2a)$$
$$k_w = \beta\, (2.85w - 10.26) + 0.612\, \alpha \qquad \text{for } 3.6 < w \leq 13 \qquad (2b)$$
$$k_w = \beta\, (5.90w - 49.91) + 0.612\, \alpha \qquad \text{for } w > 13 \qquad\qquad (2c)$$

with $\alpha = (600/Sc)^{2/3}$ and $\beta = (600/Sc)^{1/2}$.

For a given gas, Sc varies with water temperature, decreasing as the temperature increases. The dependence of Sc on sea surface temperature (SST) for DMS was presented graphically by Erickson et al. [1990] and a cubic polynomial has been fitted to that data for use in the model,

$$Sc = 3628.5 - 234.58(SST) + 7.8601\,(SST)^2 - 0.1148\,(SST)^3 \qquad\qquad (3)$$

Sea surface temperature in the Southern Ocean southwest of Cape Grim has been modelled as a periodic function with a period of one year,

$$SST = 0.5\{(SST_{max} + SST_{min}) + (SST_{max} - SST_{min}) \cos [(t - t_m)\pi/180)]\} \qquad (4)$$

where SST_{max} and SST_{min} are the maximum and minimum annual temperatures, t the time in Julian days and t_m the day on which the maximum temperature occurs. From air temperature data recorded at Cape Grim, the maximum and minimum sea surface temperatures are estimated to be 14.6°C and 9.4°C, respectively, with the maximum occurring on day 17. Thus from (3), Sc varies from a value of 1522 (at 14.6°C) to 2022 (at 9.4°C).

Photosynthesis

The original GMSK model formulation assumed that phytoplankton growth was not limited by either light or temperature, an appropriate simplification for predictions on the time scale of a bloom episode but not valid when seasonal variability is to be investigated. Assuming that both availability of light and the ambient temperature can effect the specific nutrient uptake rate, μ, at any given time - the multiplicative model [19],

$$\mu = V_N R_L R_T \qquad (5)$$

where V_N (h^{-1}) is the nitrogen-specific nutrient uptake rate following the Michaelis-Menten form given in the original GMSK formulation, and R_L and R_T are dimensionless light and temperature limitation coefficients, respectively.

Field measurements suggest nitrate levels in the Subantarctic Southern Ocean are typically very high in late winter 11.5 μM and drop to 2.1 μM by the end of the austral summer. The nitrate half-saturation concentration K_s varies with algal species size and ambient nutrient status, however oceanic species tend to have values in the range 0.1 to 1.0 μM with higher values (>4 μM) in coastal and eutrophic waters [20]. A K_s value of 1.0μM was used in the GMSK model and in the absence of better taxonomic data has been left unchanged.

A radiation model [21] has been calibrated with meteorological data on cloudiness and surface radiation collected at Cape Grim and used to compute the incident solar irradiance at the sea surface I_s, in the range 350-700 nm (photosynthetically active radiation - PAR). The ratio of PAR to total incoming solar radiation has been estimated to be in the range 0.42 to 0.48 for mid to high latitudes [22].

In the Subantarctic ocean the mixed layer depth varies seasonally from about 100 m in summer to over 400 m in winter. The euphotic zone depth is fairly constant at around 90m for the whole year. For simplicity the euphotic zone averaged irradiance I_e has been used, assuming that light is exponentially attenuated with depth, and that the irradiance at the bottom of euphotic zone ($z = Ze$) is 1% of the surface value, thus the depth-integrated irradiance is

$$I_e = (I_s/ Ze) \int_0^{Ze} e^{-kz} dz = 0.21 I_s \qquad (6)$$

Light limitation of phytoplankton growth may be modelled in a number of ways. For example, Smith [23] has suggested,

$$R_L = P / P_{max} = (I / I_k)[1 + (I/I_k)^2]^{-0.5} \qquad (7a)$$

where P is the gross photosynthesis rate, P_{max} the maximum photosynthesis rate, and I_k the saturating irradiance /24/. Steele /25/ used a formulation which includes photoinhibition at high light intensities,

$$R_L = (I / I_{opt}) \exp\{ 1 - I/I_{opt} \} \tag{7b}$$

where I_{opt} is the light intensity at which P_{max} occurs.

There is a clear temperature effect on the growth rate of phytoplankton with the maximum rate of growth roughly doubling every 10 °C increase in sea temperature /26/. The limitation due to temperature is given by,

$$R_T = (\mu_0/\mu_{max}) \exp(0.063T) \tag{8}$$

where μ_0 is 0.84 day^{-1} and T is the temperature in degrees Celsius. For the maximum sea temperature in the study window (14.6°C), the maximal growth rate μ_{max} is 2.1 day^{-1}. By comparison, for winter sea temperature of 9.4°C, the growth limitation R_T is 0.72.

RESULTS AND DISCUSSION

Remote Sensing Data

Chlorophyll pigment concentration for the Southern Ocean southwest of Cape Grim was derived from archival Coastal Zone Color Scanner (CZCS) data. The CZCS 1978-1986 monthly composite archive was processed for the 1000 x 1000 km study window 41 - 51°S and 133.3 - 143.2 °E and the results for spatially averaged chlorophyll-like pigment concentration are shown in Figure 2a and 2b. While there is no well-defined peak in the seasonal cycle, the minimum value in the northern section occurs over August to September while in the southern half the minimum occurs in June. In both sections variance is high throughout the year except during August and September. The monthly mean pigment concentration is in the range 0.2 - 0.3 mg chl m^{-3} during the December to March period.

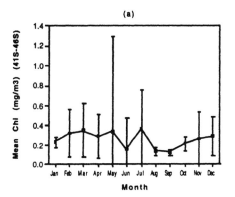
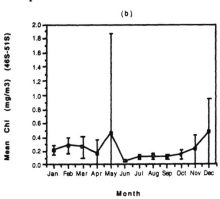

Fig. 2. Monthly average pigment concentration as derived from CZCS 1978-86 composited data (a) 41°S -46°S (b) 46°S -51°S

While there is little data on the seasonal cycles of phytoplankton species in the Subantarctic Southern Ocean, it appears that diatoms are an important component of the total flora, especially further south. McTaggart and Burton /11/ report that coccolithophores form a significant portion of the total phytoplankton community north of the Antarctic Convergence (53°S). Jacques et al. /27/ surveyed a transect in the Indian ocean sector of the Southern Ocean from 43°S to 62°S and suggested that diatoms became more dominant with increasing latitude, with species of *Chaetoceros, Rhizosolenia, Thalassiosira and Fragilariopsis* being the most important. This may cause a meridional gradient in

DMS production as coccolithophores are known to produce high levels of DMS, while diatoms are less significant producers /1/.

Model Predictions

Wind speed records collected at Cape Grim were statistically analysed and the monthly mean speed E(w) and standard deviation σ are given in Table 1.

TABLE 1 Expected monthly wind speed (m s^{-1}) at Cape Grim during 1989

	Jan	Feb	Mar	Apr	May	Jun	Jul	Aug	Sep	Oct	Nov	Dec
E(w)	11.2	9.19	8.97	8.13	8.17	7.33	7.72	7.91	8.67	9.83	9.22	10.7
σ	4.08	4.22	4.08	3.83	3.75	3.72	3.80	4.11	5.02	4.72	4.33	4.42

The wind data were used to compute the variation in transfer velocity over a year and the results plotted in Figure 3. The transfer velocity varies from a maximum value of 3.4 m d^{-1} in January to a minimum in July of 1.6 m d^{-1}. From equations (2) it is clear that k_w is directly proportional to wind speed and inversely proportional to Schmidt number, which is a minimum in January. Wind speed is a maximum in January and the combined effects of high wind and warm sea temperatures result in a maximum in k_w during summer.

Erickson et al. /28/ computed global DMS transfer velocity based on global climate model generated wind and temperature fields. Interestingly, their predictions for the Southern Ocean near Cape Grim are higher in winter (6 m d^{-1}) than in summer (3.6 m d^{-1}), although their summer prediction is very close to that presented here. Given that our findings were based on locally derived data on wind and air temperature, the winter DMS flux given by Erickson et al. may be seriously overestimated.

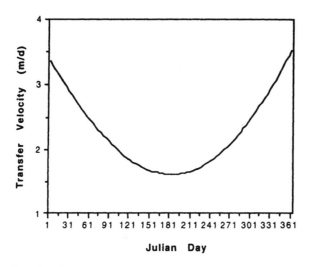

Fig. 3   Predicted variation in transfer velocity based on wind speed data collected at Cape Grim

The extended GMSK model was integrated forward in time from early October (Julian day 275) through to mid-May (Julian day 500). We have assumed that the oceanic mixed layer during spring/summer is 100 m deep. The biotic state variables were all initialised·to arbitrarily low values and the sulfur species (DMSP and DMS) were initialised to zero. For the high nitrate values typical of the Southern Ocean in early spring nitrogen is unlikely to limit phytoplankton growth so that phytoplankton growth rate is largely controlled by light and temperature.

Model predictions are known to be very sensitive to the algal cell DMSP content. Assuming that the phytoplankton community is dominated by diatoms and coccolithophores, we choose a S(DMSP):N cell ratio of 0.3 /13/. The depth-dependent model parameters were re-scaled to the 100m MLD and are given in the Appendix.

Several model runs were carried out and typical results for the growth of phytoplankton and the production of dissolved DMS are shown in Figure 4 a,b.

(a)   (b)

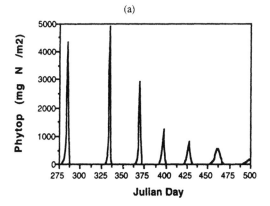
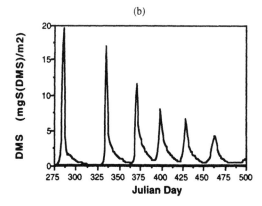

Fig. 4   Predicted variation in (a) phytoplankton biomass and (b) dissolved DMS

Gabric et al. /13/ have noted that the model predictions are very sensitive to the specification of the phytoplankton maximum nutrient uptake rate (k23) with both the timing and magnitude of DMS peaks being affected. This parameter will be species dependent and given the large area of ocean that can affect DMS measurements made at Cape Grim and the likely presence of a number of species in different stages of growth, the model predictions are clearly only an indicator of the actual cycle in DMS.

The phytoplankton growth follows a cyclical pattern with period of about 40 days and a gradual decline in the peak during autumn. DMS peaks ensue the algal peaks with a lag of about 1-2 days and follow a similar decline. DMS levels remain high for at least 10 days after phytoplankton have been completely grazed. The predicted mixed layer peak phytoplankton concentrations vary from almost 5000 to 800 mg Nm^{-2} .

In order to compare the model predictions with the pigment concentrations detected in the CZCS imagery, a currency conversion from the model units of nitrogen to chlorophyll is required. Carbon:chlorophyll ratios can vary during the bloom cycle and also across algal taxonomic group. /20/.Assuming a C:Chl ratio of 50 and a C:N of 6, the model predictions correspond to chlorophyll concentrations in the range 6 to 1 mg m^{-3}. Mixed layer peak DMS concentrations vary from 19 to 4 mg $S(DMS)m^{-2}$ which correspond to the range 6 to 1.2 nM. This is similar to the range measured by Bates et al. /8/ in the North Pacific and in the Baltic /9/.

Model predictions for the sea-to-air flux of DMS are shown in Figure 5 together with independent estimates of the monthly average flux made by Ayers et al. /29/ using the Cape Grim atmospheric measurements and data on the rate of gaseous DMS removal through reaction with the hydroxyl radical. Given the limitation that our model assumes a biologically homogeneous ocean, the correspondence between the two sets of predictions is encouraging. The flux increases toward a maximum at the end of the year and then decreases in autumn which is identical to the trend in atmospheric DMS at Cape Grim.

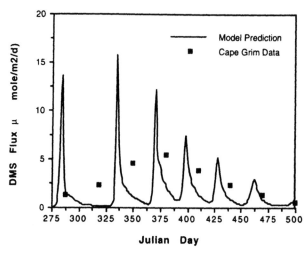

Fig. 5   Model predictions for DMS flux

It should be noted that the model predictions are based on the assumption of a spatially uniform study region. Clearly this is a gross simplification given the extent of ocean which will affect DMS measurements at Cape Grim. The phytoplankton population over such a large area will likely be heterogeneous, implying different cell DMSP content, and in different stages of the growth cycle. Thus, while the flux predictions display a similar periodicity to the dissolved DMS concentration, the effect of heterogeneity in the phytoplankton assemblage and asynchronous blooming would be to fill in the troughs and give a more uniform flux over time as is indeed found in the experimental record.

APPENDIX

Definition of Model State Variables and Fluxes

X1	phytoplankton , mg Nm^{-2}
X2	bacteria, mg Nm^{-2}
X3	zooflagellates, mg Nm^{-2}
X4	large protozoa, mg Nm^{-2}
X5	micro and mesozooplankton, mg Nm^{-2}
X6	dissolved inorganic nitrogen, mg Nm^{-2}
X7	dissolved DMSP, mg S(DMSP) m^{-2}
X8	dissolved DMS, mg S(DMS) m^{-2}
F12	bacterial decomposition of phytoplankton, mg $Nm^{-2}d^{-1}$
F14	ingestion of phytoplankton by large protozoa, mg $Nm^{-2}d^{-1}$
F15	ingestion of phytoplankton by micro-mesozooplankton,mg $Nm^{-2}d^{-1}$
F17	excretion of DMSP by phytoplankton, mg S(DMSP) $m^{-2}d^{-1}$
F18	excretion/metabolism of DMS by phytoplankton, mgS(DMS)$m^{-2}d^{-1}$
F1W	sedimentation of phytoplankton below euphotic zone, mg $Nm^{-2}d^{-1}$
F23	ingestion of bacteria by zooflagellates, mg $Nm^{-2}d^{-1}$
F26	excretion of DIN by bacteria, mg $Nm^{-2}d^{-1}$
F34	ingestion of zooflagellates by large protozoa, mg $Nm^{-2}d^{-1}$
F36	excretion of DIN by zooflagellates, mg $Nm^{-2}d^{-1}$

F45	ingestion of protozoa by zooplankton, mg $Nm^{-2}d^{-1}$
F46	excretion of DIN by large protozoa, mg $Nm^{-2}d^{-1}$
F47	excretion of DMSP by large protozoa, mg $S(DMSP) m^{-2}d^{-1}$
F56	excretion of DIN by micro and mesozooplankton, mg $Nm^{-2}d^{-1}$
F57	excretion of DMSP by zooplankton, mg $S(DMSP) m^{-2}d^{-1}$
F5W	sedimentation and export of zooplankton, mg $Nm^{-2}d^{-1}$
F61	DIN uptake by phytoplankton, mg $Nm^{-2}d^{-1}$
F62	DIN uptake by bacteria, mg $Nm^{-2}d^{-1}$
F72	DMSP biodegradation and uptake by bacteria, mg $S(DMSP) m^{-2}d^{-1}$
F78	Conversion of DMSP to DMS in water column, $mgS(DMSP)m^{-2}d^{-1}$
F82	DMS uptake/consumption by bacteria, mg $S(DMS) m^{-2}d^{-1}$
F8W	DMS photo-oxidation, adsorption/sedimentation, $mgS(DMS)m^{-2}d^{-1}$
F8A	DMS ventilation to atmosphere, mg $S(DMS) m^{-2}d^{-1}$

*F_{ij} is the flow between compartments X_i and X_j (A is atmosphere; W is water column)

Model Equations

$dX1/dt$ = F61 - F12 - F14 - F15 - F1W
$dX2/dt$ = F12 + F62 - F23 - F26 + α_1 F82 + α_2 F72
$dX3/dt$ = F23 - F36 - F34
$dX4/dt$ = F34 + F14 - F45 - F46
$dX5/dt$ = F45 + F15 - F56 - F5W
$dX6/dt$ = F26 + F36 + F46 + F56 - F61 - F62
$dX7/dt$ = F17 + F47 + F57 - F78 - F72
$dX8/dt$ = F18 + β F78 - F 82 - F8W - F8A
with
F12   = k1 X2 [1 - exp(-k2 X 1)]
F14   = k3 X1 X4
F15   = k4 X1 X5
F17   = k5 X1 γ
F18   = b g k6 X1
F1W   = k7 X1
F23   = k8 X3 [1 - exp(-k9 X2)]
F26   = k10 X2 + k11 [F62 + F12]
F34   = k12 X3 X4
F36   = k13 X3 + k14 F23
F45   = k15 X4 X5
F46   = k16 X4 + k17 [F34 + F14]
F47   = k18 X4 γ
F56   = k19 X5 + k20 [F15 + F45]
F57   = k21 X5 γ
F5W   = k22 F56 + k32 X5
F61   = k23 X1 [1 - exp(-k24 X6)]
F62   = k25 X2 [1 - exp(-k26 X6)]
F72   = k31 X7
F78   = k27 X7
F82   = k28 X8
F8W   = k29 X8
F8A   = k30 X8

Currency conversion factors α_1 and α_2 are given by 0.15 and 0.37, respectively, and γ is the S(DMSP):N ratio for phytoplankton (species dependent).

Model Parameters for a 100-m mixed layer

	Value	Pathway*	Units
k1	4.5	P-B	day^{-1}
k2	4.6e-4	P-B	m^2 mg N^{-1}
k3	2.6e-3	P-LP	m^2 mg N^{-1} d^{-1}
k4	1.2e-3	P-Z	m^2 mg N^{-1} d^{-1}
k5	0.01	P-DMSP	day^{-1}
k6	0.0085	P-DMS	day^{-1}
k7	0.15	P Sinking	day^{-1}
k8	17	B-F	day^{-1}
k9	1.38e-3	B-F	m^2 mg N^{-1}
k10	0.07	B-N	day^{-1}
k11	0.63	B-N	...
k12	0.0156	F-LP	m^2 mg N^{-1} d^{-1}
k13	0.05	F-N	day^{-1}
k14	0.65	F-N	...
k15	1.2e-3	LP-Z	m^2 mg N^{-1} d^{-1}
k16	0.05	LP-N	day^{-1}
k17	0.65	LP-N	...
k18	0.01	LP-DMSP	day^{-1}
k19	0.05	Z-N	day^{-1}
k20	0.40	Z-N	...
k21	0.01	Z-DMSP	day^{-1}
k22	0.15	Z sinking	...
k23	0.9	N-P	day^{-1}
k24	5e-4	N-P	m^2 mg N^{-1}
k25	0.9	N-B	day^{-1}
k26	9.24e-3	N-B	m^2 mg N^{-1}
k27	0.5	DMSP-DMS	day^{-1}
k28	0.95	DMS-B	day^{-1}
k29	0.27	DMS photo-ox	day^{-1}
k30	see text	DMS - atmos	day^{-1}
k31	1.0	DMSP-B	day^{-1}
k32	0.05	Z export	day^{-1}

REFERENCES

1. M.D. Keller, WK Bellows and RL Guillard. Dimethyl Sulfide production in marine phytoplankton. In, *Biogenic sulfur in the environment*, Saltzman ES and Cooper WJ (eds), American Chemical Society, Washington, DC , 1989.

2. S.M. Turner, G Malin , PS Liss , DS Harbour and PM Holligan. The seasonal variation of dimethylsulfide and dimethylsulfoniopropionate concentrations in nearshore waters. *Limnol. Oceanogr.*, 33(3), 364-375, 1988.

3. RL Iverson, FL Nearhoof and MO Andreae. Production of dimethylsulfonium and dimethylsulfide by phytoplankton in estuarine and coastal waters. *Limnol. Oceanogr.*, 34(1), 53-67,1983.

4. S. Belviso, S.-K. Kim, F. Rassoulzadegan, B.Krjka, B.C.Nguyen, N.Mihalopoulos, and P.Buat-Menard. Production of dimethylsulfonium propionate (DMSP) and dimethylsulfide (DMS) by a microbial food web. *Limnol. Oceanogr.,* 35(8): 1810-1821, 1990.

5. T.S. Bates, B.K. Lamb, A. Guenther, J. Dignon, and R.E. Stoiber. Sulfur emissions to the atmosphere from natural sources, *J. Atmos. Chem.,* 14, 315-337, 1992.

6. R.J. Charlson, J.E. Lovelock , M.O. Andreae and S.G. Warren. Oceanic phytoplankton, atmospheric sulphur, cloud albedo and climate. *Nature,* 326, 655-661, (1987).

7. G.P. Ayers, JP Ivey and RW Gillett. Coherence between seasonal cycles of dimethylsulphide, methanesulphonate and sulphate in marine air. *Nature,* 349, 404-406, 1991.

8. T.S. Bates, JD Cline , RH Gammon and SR Kelly-Hansen. Regional and seasonal variations in the flux of oceanic dimethylsulfide to the atmosphere. *J Geophys. Res.,* 92, C3, 2930-2938, 1987.

9. C. Leck, U Larsson , LE Bagander , S Johansson and S Hajdu. DMS in the Baltic Sea - Annual variability in relation to biological activity. *J Geophys. Res.,* 95 (C3), 3353-3363, 1990.

10. B.C. Nguyen , N Mihalopoulos and S Belviso. Seasonal variation of atmospheric dimethylsulphide at Amsterdam Island in the Southern Indian Ocean, *J. Atmos Chem.,* 11, 123-141, 1990.

11. A.R. McTaggart, and H Burton. Dimethyl sulfide concentrations in the surface waters of the Australasian Antarctic and Subantarctic oceans during an austral summer. *J. Geophys. Res.,* 97, 14407-14412, 1992.

12. M.O. Andreae, and W.R. Barnard. The marine chemistry of dimethylsulphide. *Mar. Chem.,* 14: 267-279, 1984.

13. A.J. Gabric, Murray N, Stone L and Kohl M. Modelling the production of dimethylsulfide during a phytoplankton bloom. *J. Geophys. Res.,* 98, C12, 22805-22816, 1993.

14. F. Azam, T Fenchel, JS Gray, LA Meyer-Reil and F Thingstad. The ecological role of water-column microbes in the sea., *Mar. Ecol. Prog. Ser.,* 10, 257-263, 1983.

15. T. Fenchel, Marine plankton food chains. *Ann. Rev. Ecol. Syst.,* 19, 19-38, 1988.

16. R.P. Kiene and TS Bates. Biological removal of dimethylsulfide from seawater. *Nature,* 345, 702-705, 1990.

17. P. Brimblecombe and Shooter D. Photo-oxidation of dimethylsulfide in aqueous solution. *Mar. Chem..,* 19, 343-353, 1986.

18. P.S. Liss and L Merlivat. Air-sea gas exchange rates: Introduction and synthesis. In, *The role of air-sea exchange in geochemical cycling,* P. Buat-Menard (ed) 113-127, Reidel, Hingham, MA,1986.

19. T. Platt, Denman,K.L. and Jassby, A.D. Modeling the productivity of phytoplankton, *in The Sea,* Vol.6, E.D.Goldberg (ed.) New York, Wiley, 807-856, 1977.

20. T.R. Parsons , Takahashi M and Hargrave B. *Biological Oceanographic Processes.* 3rd Ed, Pergamon Press, 1984.

21. T.D.Brock, Caculating solar radiation for ecological studies, *Ecol Model.,* 14, 1-19, 1981.

22. K.S.Baker and Frouin R.   Relation between photosynthetically available radiation and total insolation at the ocean surface under clear skies. *Limnol. Oceanogr.* , 32(6), 1370-1377, 1987.

23. E.L. Smith. Photosynthesis in relation to light and carbon dioxide, *Proc. Nat. Acad. Sci. Wash.*, 22, 504-511, 1936.

24. J.F. Talling. Photosynthetic characteristics of some freshwater plankton diatoms in relation to underwater radiation, *New Phytol.*, 56, 29-50, 1957.

25. J.H. Steele. Environmental control of photosynthesis in the sea. *Limnol. Oceanogr.*, 7, 137-150, 1962.

26. R.W. Eppley. Temperature and phytoplankton growth in the sea, *Fish. Bull.*, 70, 1063-1085, 1972.

27. G. Jacques , Descolas-Gros C, Grall JR amd Sournia A. Distribution du phytoplancton dans le partie antarctique de l'Ocean Indienne en fin d'ete, *Int. Rev. Res. Hydrobiol.*, 64, 609-628, 1979.

28. D.J. Erickson, Ghan SJ and Penner JE. Global ocean-to-atmosphere dimethyl sulfide flux., *J. Geophys. Res.*, 95, D6, 7543-7552, 1990.

29. G.P. Ayers, Ivey JP, Bentley ST and Forgan BW. Dimethylsulphide in marine air at Cape Grim, 41°S, *Tellus* (in press)

Pergamon

Adv. Space Res. Vol. 18, No. 7, pp. (7)129–(7)132, 1996
Copyright © 1995 COSPAR
Printed in Great Britain. All rights reserved
0273–1177/96 $9.50 + 0.00

0273–1177(95)00955–8

QUASISTATIONARY AREAS OF CHLOROPHYLL CONCENTRATION IN THE WORLD OCEAN AS OBSERVED SATELLITE DATA

A. P. Shevyrnogov, G. S. Vysotskaya and J. I. Gitelson

Institute of Biophysics (Russian Academy of Sciences, Siberian Branch), Academgorodok 660036 Krasnoyarsk, Russia

ABSTRACT

To estimate the seasonal progress of the production process over the planet and its long-standing trend it is important to measure not only the spatial distributions of pigment that represent the rates of photosynthesis but also their time variability. Anthropogenic impact on natural complexes can be efficiently estimated by satellite observations of phytopigment dynamics. This study presents CZCS satellite data processed to reveal spatial inhomogeneity in the seasonal course of chlorophyll concentration in the world ocean on a global scale. Areas with quasistationary and non-stationary seasonal changes in chlorophyll concentration are revealed for the 7.5 years period of CZCS operation. Areas in the world ocean with maximum absolute and relative differences in chlorophyll concentration during different seasons were found. Results are illustrated with a chart of areas in the Atlantic Ocean discerned by dynamic parameters.

SPATIAL INHOMOGENEITY OF SEASONAL PROGRESS OF CHLOROPHYLL CONCENTRATION IN THE WORLD OCEAN

For oceanologists the basic challenge is to reveal long-standing trends in the variability of biological productivity of water areas as well as local dynamics of phytoplankton concentration that are dependent on anthropogenic impact and natural processes. To determine these dynamics it is necessary to reveal quasistationary areas (QSA) that are fit for systematic contact and remote measurements. Of great interest are areas with elevated variability which are generally associated with frontal zones, upwelling boundaries, etc. /1/.

It is apparent that small spatial and temporal resolution prevent the use of the standard biological techniques (extraction, for example) to estimate the quantity and biological composition of the oceanic plankton. Long-standing satellite data concerning chlorophyll distributions in the superficial layer of the ocean obtained with the CZCS sensor make possible a new mode of thought about the dynamics of biological processes in oceanic ecosystems. Variability of chlorophyll concentration and its seasonal progress, in particular, seem to be one of efficient parameters that reveal this dynamics /2/.

Methods

The method developed at the Institute of Biophysics (Russian Academy of Sciences, Siberian Branch) for revealing quasistationary and enhanced biological activity areas in the ocean uses moving variance as applied to spatial distribution of seasonal variability of chlorophyll concentration (the satellite data donated to the Institute of Biophysics by NASA). Variance is calculated over elementary squares. Let $A_{k,i,j} = \{a_{m,n}\}$ be a set of pixel values which are inside of a square with a center (i,j) for season K. $D_{k,i,j}$ is a standard deviation for $A_{r,i,j}$. We shall call "moving variance" $V_{i,j}$ in point (i,j) an average of $D_{1,i,j}, D_{2,i,j}, D_{3,i,j}, D_{4,i,j}$. The optimums for such elementary sites were two sizes of squares. One was sized 11×11 pixels - 121 elements, with the square side 42 miles and the other was 5×5 pixels, (25 elements), with the square side of 19 miles. Each of these estimates was effective for the areas to be investigated. Good for large areas such as the Northern, Central, Southern Pacific, Atlantic and Indian oceans - were the 11×11 squares. For coastal and inland seas the most efficient site was 5×5 squares.

A. P. Shevymogov *et al.*

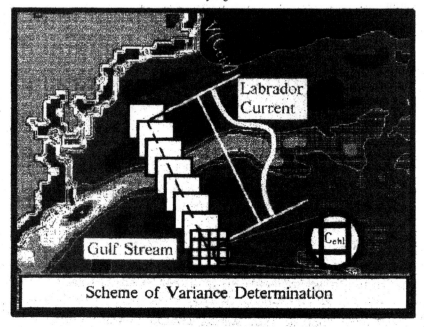

Fig. 1. Determination of moving variance of chlorophyll concentration

The non-optimality measure for the estimates applied was either degradation of the areas highlighted or excessive splitting of the image. Correction introduced for the shape of the statistical distribution of the original data in chlorophyll concentration enabled us to discern locations with high pigment concentration such as areas of upwelling, river estuaries and large cities; considerable variations of absolute values notwithstanding.

Figure 1 shows the diagram of deriving moving variance using the intersection of the frontal zone between the Labrador Current and Gulf Stream as an example. Even though the seasonal progress of dynamics in chlorophyll concentration is different and so are the absolute chlorophyll concentrations, the variance on both sides of the frontal zone is small. The variance drastically increases at the intersection of the interface between the waters with different seasonal dynamics of chlorophyll concentration. The value of the moving variance is shown in the chart with different optical density.

<u>Results</u>

Figure 3 shows a chart of spatial distribution of moving variance of chlorophyll concentration derived by the method described above. A comparison with the chart of chlorophyll concentration spatial distribution (Figure 2 - Atlantic, winter in the northern hemisphere) shows the difference in spatial distribution of chlorophyll concentration and its moving variance in all regions of the Atlantic.

The physical sense of the moving variance spatial distribution chart is to highlight the areas with similar seasonal progress of chlorophyll concentration and slow changes of absolute values, by coloring zones corresponding to small variance values. In this manner the chart of the Atlantic Ocean shows QSA in the form of limited areas of different size and shape. These probably results from differences in the nature of production occurring there and the pigment destruction processes, both dependent on specific biogeographical conditions.

Emergence of homogeneous areas is quite different. Spatial distribution of phytoplankton and accordingly of chlorophyll concentration is primarily affected by concentration of nutrients biogenous elements, temperature, light conditions determining intensity of biosynthesis, etc. All these circumstances are determined by the ocean dynamics and the latitude under investigation.

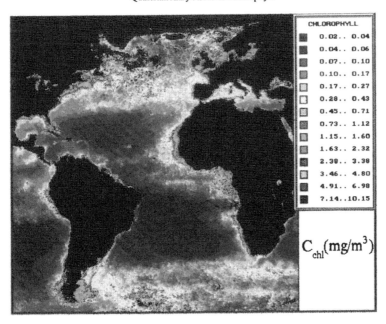

Fig. 2. Chlorophyll concentration in superficial Atlantic waters (winter in the northern hemisphere)

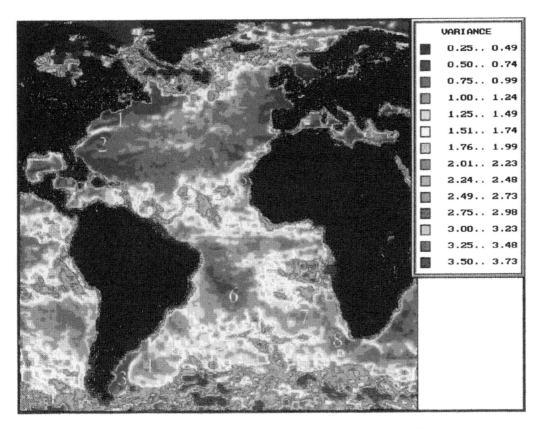

Fig. 3. Moving variance of chlorophyll concentration in the Atlantic Ocean (Quasistationary zones are revealed with this method, see explanation in the text).

Developed method allows us to see remarkable QSA in different areas of world ocean. In Figure 3 the northern Atlantic clearly exhibits QSA on both sides of the front between the Gulf Stream and Labrador Current (1). Differences in the seasonal dynamics of chlorophyll concentration are due to different of water masses and temperature conditions.

The southern Atlantic exhibits an analogous picture, with QSA manifested on both sides of the front dividing the warm waters of the Brazil Current (4) and the cold Antarctic waters of the Falkland Current (3).

An example of a QSA in a coastal upwelling area is that at Cap Blanc, northwest Africa (5). In this area affected by trade winds, waters are observed to upwell all year round with accordingly high biomass of phyto- and zooplankton [2]. Analogous pictures are observed in other coastal upwelling areas. It is noteworthy that QSA are reliably discerned, depending on the nature of seasonal chlorophyll variability irrespective of the absolute value of chlorophyll concentration.

Most impressive is the picture of coincidence of QSA (6, 7, 8) with the divergence line in the southern anticyclone subtropical turnover (formed by Southern trade wind, Brazil, Southern Atlantic and Benguela currents) between 20° and 40° southern latitudes /3/.

Measurements and processing covering almost the entire surface of the global ocean produced some manifestations of QSA that defy interpretation, in particular the small scale and high latitude QSA's. Of interest were some observed anomalies and local QSA in discharge areas of large rivers and large industrial centers which may be due to anthropogenic impact.

REFERENCES

1. *Oceanology. Ocean Biology.* Vol. 1. "Biological structure of the ocean" (Ed. M.E. Vinogradov). - M., "Nauka", 1977.

2. J.I.Gitelson, A.P.Shevyrnogov, S.L.Molvinskikh, V.V.Chepilov Determination of Photosynthetic Pigments in Aqueous Ecosystems, in: *SCOPE/UNEP Sonderband*, Hamburg, Heft 66, 1988, s. 331-340.

3. R.P.Bulatov, Circulation of Atlantic ocean waters in different space-and-time scales, in: *Oceanologic research.* M.: "Nauka", 1971, № 22, p. 7-93.

 Pergamon

Adv. Space Res. Vol. 18, No. 7, pp. (7)133–(7)139, 1996
Copyright © 1995 COSPAR
Printed in Great Britain. All rights reserved
0273–1177/96 $9.50 + 0.00

0273–1177(95)00956–6

AN APPLICATION OF AERIAL REMOTE SENSING TO MONITOR SALINIZATION AT XINDING BASIN

Qiao Yu-liang

Shanxi Agricultural Remote Sensing Application Institute, Taiyuan, China

ABSTRACT

In this paper, a method to interpret the high, mid, low salinized ploughland and the salinized wasteland using comprehensive aerophoto interpretation principles will be described for Xinding Basin, Shanxi Province. The dynamic change of salinized soil during 7 years from 1980 to 1987 will be compared with the typical Dingxiang County. The map and data obtained, with an accuracy of more than 90%, are provided to the local government as the scientific grounds to instruct agricultural productivity.

Soil salinization is a worldwide problem. With the sharp increase in world population and modern industrialisation development, the natural resource consumption is increasing day and day, and bringing about a lack of land resource worldwide. As a kind of back-up land resource, salinized land has not only attracted the concern and study of the agricultural scientists in all countries, but also by the whole society. Shanxi is such a province in China where more than 1/3 of its total area of irrigation land is salinized. The statistics used to monitor this salinized area lack objectivity and accuracy. In 1987, the government of Shanxi Province began to investigate the salinized area of the whole province, using remote sensing technology. We selected the Xinding Basin in central Shanxi as the test district to perform the aerial remote sensing investigation, and, at the same time, studied the salinization dynamic change on the Dingxiang County used as the typical district.

XINDING: THE TEST DISTRICT

The Xinding Basin is in the Futuo river valley, between the Wutai and Yunzhong mountains. It is 173 km long, 12.5-42.5 km wide and 2157 square kilometres in total area. It includes the plain and gentle hillock lower than 1000m of the Xinzhou, Dingxiang, Yuanping, Dai, Fanshi County etc, and belongs to the warm-temperature zone, with an average air temperature of 8-9°C, an accumulated temperature (>10°C) of 3000-3500°C, and frost-free period of 160 days. Its annual rainfall is 450mm, and the evaporation capacity is 3-4 times that. Influenced by the climate and topography characteristics, there are 2 reasons for salinized land to form in this district. The first one is the dry climate and high mineralisation of underground water, creating circumstances suitable for salt accumulation as the soil salt capacity is increasing gradually. The second is because of the unreasonable irrigation that causes the underground water level to rise, and makes the soil become secondary salinization.

MAIN TECHNICAL METHODS

Salinized Land Interpretation

The color infrared reversal film was photographed during the first ten-day period of September, 1987, with a scale of 1:19000, and a photochromy width of 24*24 cm. The crops were just mature but not being harvested during the time of photography. In order to understand the image characteristics, we arranged the para-synchronous field sample investigation during the photography period. To obtain the general characteristics of the whole district and carry the comprehensive interpretation principles, we set all the aerophotographs of the whole test district along the first flight. After that, we analysed these photochromes against the relevant thematic maps and written materials, and delineated the study area, interpreted aerophotographs piece by piece under the stereo scope, and drew out the interpretation sketch. Finally, after the field confirmation we took the procedure of area measuring and map making etc.

Salinized Land Dynamic Change Analysis

Using the color infrared aerophotographs from 1987, with a scale of 1:19000 and panchromatic black and white aerial photos from 1980, with a scale of 1:17000 for Dingxiang County, the salinized land dynamic change analysis could be made since these two temporal aerophotos objectively record the salinized land distribution. The objective salinized land dynamic change analysis could be made since this. By the interpretation, we could draw the outline of salinized land in each stage. Then analysed comparatively the interpretation results from the aerial photo (1980) were compared to those from the color infrared photo (1987). The relative increase or decrease in the salinized land polygons increased or decreased in subsequent photographs was estimated and transformed to the 1:10000 topographic map. Finally, we could get the salinized land dynamic change data and a map of the change in Dingxiang County from 1980 to 1987. Besides the interpretation of salinized land dynamic change for those two years, we also analysed the dynamic change in landuse.

DIFFERENTIATION OF SALINIZATION LAND TYPE

Influenced by the topography, geomorphology, underground water and human activity, the soil salinization is also reflected by the degree to which it does harm to crop growth, so the natural vegetation and crop productivity can be used as an indicator of salinizaion. Distinct soil types with different salinization extent have different shortage percentage in crop seeding and growth level, and different characteristics in the color infrared photos. Additionally, according to the salt capacity and salt type, we can also distinguish the salt salinization. There are 4 main types of salinized land in Xinding Basin:

Sulfate-saline Soil: It is mainly distributed on the first-stage terrace or over the flood meadow near the two banks of mid-lower reaches of the Futuo river, and is mostly the lightly salinized soil.

Chloride-saline Soil: also distributed primarily on the first-stage terrace and the alluvial plain and low ground of the banks of the alluvial Futuo river reaches, and is a typical mid salinization soil.

Soda-saline Soil: severely salinized soil that is distributed on the first-stage terrace low ground of the Futuo, Yunzhong and Muma rivers.

Mixed-saline Soil: It mainly distributes on the low ground land alluvial plain in the basin center, which has a different damage.

Consequently, we can distinguish the non-salinized land, light salinized land, mid salinized land, severe salinized land and salinized wasteland from the infrared photos. The differentiation standards are listed below in Table 1.

TABLE 1   The extent of salinized land differentiation of Xinding Basin

Landscape description in field investigation			The test results of soil samples
Type of land	Crop cover extent (%)	Area of the salty basin in plough land (%)	Whole salinity in plough layer (%)
Salinized wasteland	difficult for crop growing	> 80	> 0.8
Severe salinized land	20-30	60-80	0.6 - 0.8
Mid salinized land	30-70	20-60	0.3 - 0.6
Light salinized land	> 70	10-20	0.1 - 0.3
Non salinized land	100	0	<0.1

INTERPRETATION OF SALINIZED LAND USING INFRARED PHOTOS

Using our interpretation of the salinized land from the infrared photos, we have completed a comprehensive analysis of the salinized land types according to the several kinds of factors, such as the crop type, growth circumstances, irrigation installation and the topographic site, etc.

The coarse food grains, such as sorghum, maize, wheat, cereals, and beans etc, are the main crops in Xinding Basin. In general, the wheat multiple-cropping land and the maize-sorghum inter plant land are non-salinized land. This type of crop and its planting method is one of the important marks of salinized land discrimination. Whereas, the non inter-planting land of maize, sorghum, beans and so on, are mostly with different salinization extent. According to the crop plant principle, we can generally distinguish between the cultivated land with or without salinization by the crops that occur. Using this, the difference of image color, tone, saturation and uniform extent of the same crop type as interpretation marks, assisted on the image shape, texture and distribution site, we can classify the levels of salinization extent. The salinized land interpretation could be related to the crop growth.

Because of the degree of vegetation and chlorophyll, the crop with good growth level has a intense "infrared steep slope effect" in its reflection spectrum, which leads to a bright red color on the aerophotographs. The higher the crop growth level, the denser the red color, and the higher the saturation extent. Just because of the different crop type and plant time, the image of some crops which are going to mature changes from red high saturation to orange red or orange yellow. Additionally the land with a high growth level has generally a high level of cultivation management, and accordingly a clear and tidy land outline. Because of the effect of inter-planting crops, there are also some parallel lines with 0.1mm interval within the land image. This kind of land generally belongs to the non-salinization land.

The red color of the mid-growth level land image is obviously less intense than the fertile land, it is mostly orange yellow, with an obviously lower saturation and non-uniform extent, there are even some tiny shallow non-uniform spots on the aerophotographs. According to the para-synchronous field investigation, this kind of land has more than 70% crop cover, and belongs to light salinization land type.

The reflection spectrum of the land with low crop growth level has an even lower "infrared steep slope effect". The image color of this kind of land is yellow, gray yellow, or gray. The tone of the land image is very non-uniform; there are even some partly-colored salty basins on the land image, this is the reflection of the uneven seeding and growth level. Generally, this kind of land is mid salinization type with a crop cover between 30% and 70%.

The crop cover extent of the severe salinized land is about 20% and 30%, or even lower. Although some water conservancy projects, and soil amelioration installations appear in the aerophotographs, there are low levels of artificial herbage, and the land image has a very pale red color and inhomogeneous tone. Since this kind of salinized land has no regular outline or nor can the drainage system be seen(except for some salt-tolerant grass with uneven density), it makes the land image tone very inhomogeneous. In addition, the salinized wasteland generally distributes around the salt concentration center or the edge of the alluvium where there is a bad hydro-geological condition, so it is very easily interpreted.

Although the type of the salinized land is related to the crop growth in some degree, it is not all the time. Since the crop growth level is related not only to the soil salinization extent directly, but also to the production management (such as the irrigation, fertilizer etc). Because of the unsuccessful management of fertilizer application, irrigation, drainage etc, there is also bad crop growth and lack of seeding even in non-salinized land. Conversely, with good land management, the salinized land may have good crop growth also.

The salinized land with no crop cover could be interpreted directly as a consequence of the salt concentration of the soil surface, which has a different density and inhomogeneous extent of the white color tone on the infrared aerophotographs. Light and gray colors are the sign of lightly salinized land , especially when the gray spots are increasing in size, even joining each other, it means a more severe degree of salinization. Between these two is the mid salinized land. The salinized wasteland is similar to the severe salinizedland in the image color and tone, and has generally no regular outline which is the characteristic of the cultivated land on the image, and has no artificial water conservancy and soil reform projects on it as well. There are some salt plants around the water area, their shadow present a different yellow-orange color on the aerophotos. In Xinding there is sulfate saline soil as the dominant soil, besides a certain content of chloride-saline and soda-saline soil in some areas. The soda-saline soil has a special image of inhomogeneous yellow-green spots or tiny tracks against a white background on the infrared aerophotos. The salinized soil that contains mainly chloride, especially chloride-magnesium, is easy deliquescent, and even dissolves the organic soil matter, hence, low down the spectrum reflectivity in infrared band of the soil surface, so has a blue green color in the aerophotos.

In a word, to discriminate the salinized land and their type from the infrared aerophotos we must take account of a number of factors.

RESULT ANALYSIS

Interpretation of the infrared aerophotos (September,1987), there are 27800 ha. of total salinized land in Xinding Basin, and the areas of the light, mid, severe and salinized wasteland are 8867,7533, 6533 and 4867 ha respectively. The salinized land area within irrigated land is 25867 ha; as 93% of the total salinized area.

According to the remote sensing monitoring using infrared aerophotos (1987), the salinized land area of Dingxiang County is 10400 ha., 287 ha. less than that of agricultural department statistics(1987), the difference is 2.6%, as Table 2.

TABLE 2   The salinization area comparison of the remote sensing monitoring and agricultural department statistics

Item	Total area	Light salinized land	Mid salinized land	Severe salinized land	salinized
wasteland Agricultural department statistics (1985,ha.)	10687	3700	2500	2780	1707
Remote sensing monitoring (1987, ha.)	10407	3774	2300	2840	1493
The difference (ha.)	+ 280	- 74	+ 200	- 60	+ 214
%	+ 2.6	- 2.0	+ 8.0	- 2.2	+ 12.5

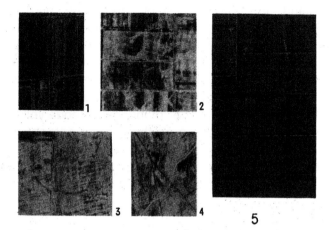

1.   Light salinized land

2.   Mid salinized land

3.   Severe salinized land

4.   Salinized wasteland

5.   Non salinized land

The salinized area decreased 280 ha between 1985 and 1987, and shows the effect of effective salinization harnessing. There is a high speed of salinization harnessing in Dingxiang County in the last two years, the first target of salinized land harnessing is to reform the mid salinized land to highproductivity crop land, then the salinized wasteland to fish pond or rice land.

This comparative analysis shows that the salinized area has decreased from 11720ha in 1980 to 10407 ha in 1987, within the 7 year's since economic system reform of the countryside, the decreasing range is 11.2% (Table 3). There are 2 reasons for this decrease. One is the decreasing underground watercapacity from

TABLE 3   The Change of the Salinized Land Area of Dingxiang
                in Remote Sensing Monitoring   (unit: ha.)

Item	Total area	Light salinized land	Mid salinized land	Severe salinized land	salinized wasteland
1980	3601	3472	3100	1548	11720
1987	2973	3100	2840	1493	10407
Difference	628	372	260	55	1313

the rain fall supply, and the increasing capacity of underground water mining, leading to the underground water to be reduced. The second reason, more importantly, is that with a greater enthusiasm and wider use of agricultural technology, the salinized land application and harnessing have been strengthened. In the total 647 ha. salinized land, there are almost 60 ha. lightly salinized land being reformed to high and stable yield land, 33 ha. mid salinized land to normal high yield land, 267 ha. severe salinized land to high yield rice land, 67 ha. to fish pond, and 33 ha. to the trees. It should also be noticed that at the same time to the great achievement of salinized land harnessing in Dingxiang County, there are about 67 ha. irrigated land to mid salinized land in some district because of the underground water level raised from the intensive flooding irrigation,this kind of reverse transformation should be attached more importance. ha. of the total salinized land in Xinding Basin, and the each area.

REFERENCE

Dai Chang Da, Land Investigation by Aerial Remote Sensing, *Agricultural Press of China, 1980.*

AUTHOR INDEX